Wine and Its Counterfeits ..

WINE

AND

I̧ŢŞ ÇOUŅŢEŖFEIŢŞ.

" A mistake is not the less so, and will never grow into a truth, because we have believed it a long time, though perhaps it be the harder to part with ; and an error is not the less dangerous, nor the less contrary to truth, because it is cried up and had in veneration by any party."—Locke.

By JAMES L. DENMAN,

20, PICCADILLY, LONDON.

1876.

LONDON :
BRISCOE AND CO., PRINTERS, BANNER STREET, FINSBURY.

CONTENTS.

HE subjoined letter from Dr. Bartlett (one of our leading analytical chemists) giving the results of his analyses of both new and matured Greek wines, was unfortunately not received until *" Wine and its Counterfeits "* had been through the press. This will account for its somewhat irregular appearance in the place it now occupies.—J.L.D.

> *Laboratory : 7, South Square,*
> *Gray's Inn, London, W.C.*
> *July 24th,* 1876.

Mr. J. L. Denman,

Sir,

I am happy to send you a condensed Report of some of the more important results obtained after a considerable number of analyses of Greek wines procured from your firm. The abstruse investigations concerning the volatile oils and ethers, resulting from prolonged but insensible fermentation, were undertaken entirely for my own purposes; and as I intend to proceed further with these interesting experiments, with a view to publication when complete, I can only allude to these compounds discovered as forming those delicate yet powerful bouquets almost peculiar to some of the Greek wines when fully matured.

I will first take a few samples of comparatively new wines, which were composed as follows:

WINES RECEIVED AUGUST, 1874.	ST ELIE	KEPHISIA, WHITE.	NOUSSA.	LACHRYMA CHRISTI
Specific gravity of wine	991·51	995 77	991 99	1083·93
Per centage of proof spirit	26 16	22 81	24·14	14·12
Total extract	2 85	2·63	3·39	24·76
Ash	·22	·23	·22	·47
Fixed acid as tartaric	·31	·42	·71	·37
Volatile acid	·18	·17	·12	·10

The St. Elie and White Kephisia already contained appreciable quantities of fragrant ethers, but not enough to make an exact estimation.

It is not necessary to give particulars of the more ordinary determination of the analyses of the splendid samples of Greek wines old in bottle, which were devoted to the infinitely more tedious analysis by fractionation. Thera, St. Elie, and White Kephisia of considerable age (10 or 12 years), yielded evidence of the most complex compound ethers. Minute quantities of tartaric ether, united with several alcohol radicles, are combined with aceto-ethylic ether, and what are sometimes erroneously termed essential oils—namely, the butyric, caproic, caprylic, and propylic ethers.

The difference between new wine containing these matters in a fixed condition and the magnificent old wines in which they are developed in a volatile form, is easily proved to the nose and palate; but by analysis it can be understood from the fact of the new Greek wines containing a rough average of about ·04 of the fixed to ·02 of the volatile, while the 10 years old wines gave ·03 of the volatile to ·01 of the fixed. Neither French nor German wines possess above half these quantities. This is accounted for by the greater sweetness and full maturity of the grapes when gathered in the Greek vineyards, yielding a far larger quantity of glucose and other constituents giving the highest fulness of body.

I ought perhaps to append to the more strictly scientific and chemical description of your Greek wines, some few general remarks which may be easier of comprehension.

I must premise that in my experience of wines produced at the high temperatures of Greece and other countries where the highest flavours and most robust wines are alone obtainable, the formation of a small quantity of acetic acid cannot always be guarded against. This is of little or no consequence to the soundness of the wine, so long as the acetic fermentation is effectually checked by the abundant formation of natural alcohol. In other words, the growth of the *mycoderma aceti* may be, and generally is, arrested by the vigorous growth of the *saccharomycis ellipsoideus*, the alcoholic ferment or yeast plant of wine.

The prevention of further acetous change in strong natural wines can only be assured when the greatest care is taken in the pressing and during fermentation. Some slight variation in the proportion of acetic acid so formed must be expected, and in fact must always be found; but, as I have explained in my lectures on "Food and its effects," published under the title of * "Cup and Platter," neither the presence of natural acid, nor the variation in its quantity (in moderation) can be considered objectionable, but the reverse. Sir James Eyre, M.D., states, in his work on the "Stomach and its difficulties," "that organ, when healthy, *enjoys* vegetable acids, the citric and acetic, especially the

* H. S. King and Co., Cornhill, E.C.

latter, the patient may train it by degrees to return to the moderate use of ascescent drinks." So, then, it is considered very desirable to health to find a moderate proportion of such acids in wine. I may here observe that the finer flavours and bouquets cannot be developed without the acetic ethers which are incompatible with the absence of the natural acids.

Tartaric acid in wines is, however, frequently mistaken by the uneducated palate for what is termed " acidity," but unless there is a sufficiency of tartaric acid in the wine, it cannot yield that refreshing quality which is, at least, as valuable as its diffusive stimulus. Fortification precipitates the natural tartrates, so does plastering ; the consequence is that flat, insipid, spirituous mixtures, are easily brought forward as "*free from acidity*," and as such are not in *reality wines at all.*

The Greek wines, without being unduly acid, have not had their natural components removed by added alcohol or by plastering. Neither, I am happy to say, are they " deplastered " by the dangerous process lately patented by two separate chemists, although the *modus operandi* has been known to me as having been in use for the last two years. I can only deprecate, as being worse than the plastering, any recourse to such after treatment for the removal of the salts of potassium left in thereby, which, by the way, are *not* those of plaster.

If any words of mine can add to the genuine and superior character which distinguishes the better class wines of Greece, I shall be pleased to have the opportunity of tendering my testimony to that effect.

Faithfully yours,

H. C. BARTLETT, Ph.D., F.C.S.

WINE, AND ITS COUNTERFEITS.

"A mistake is not the less so, and will never grow into a truth, because we have believed it a long time, though perhaps it be the harder to part with ; and an error is not the less dangerous, nor the less contrary to truth, because it is cried up and had in veneration by any party."—Locke.

IT is not well to prefer novelty merely for the sake of change ; neither should one be so prejudiced in favour of the antique as to discredit new things because they are new. There will, doubtless, always be found advocates of the old established ways and usages, inasmuch as it is easier to do what habit suggests, than to think out new thoughts, and to pursue the best wheresoever it may be obtained ; so also there will always be searchers and explorers, for to many minds activity and endeavour are the chief delights of existence. In due course, however, there at length comes a time when that which was novel is so no longer, and when, in order that it may survive the period of infancy, it must show its right to endure—its *raison d'être*. From this point of view I propose to make a few inquiries as to some changes which have taken place in the public taste during the past 15 years, with the purpose of ascertaining whether there has or has not been, during that time, a general advance in opinion upon the lines I then assisted in laying down, and whether the theories which were then comparatively novel and unheard of have survived the ordeal of time.

It is, I believe, quite impossible for any ordinary observer to deny that the attacks which have been carried on all those years, more or less desultorily, against the Port and Sherry citadels, have been deservedly successful. These old compounds have been tried and found wanting. They have been compared disadvantageously with every pure new wine; until at length Spain and Portugal, which have well and sturdily battled for their long-established and sole privilege to mix John Bull's very peculiar draughts, are obviously disposed to mitigate his dose, so as to make it a little more like wine, and a little less like a liqueur.

When, in 1861, I began to show that Port and Sherry were,—from being so heavily loaded with spirit and sugar, and from being beplastered and incompletely fermented—decidedly not wholesome—my statements were in many places received with amazement and incredulity; and there were not a few who held the opinion that not France and Greece, but Spain and Portugal produced true wine. Wine merchants with large stocks, gouty old gentlemen with well-filled cellars of Port and Sherry, properly cobwebbed and sealed, heard my impeachment of them with supreme disgust, and sometimes exhibited almost as much natural horror and indignation as an idolater might at the anathema of a missionary. In a word, our two old wine-deformities were undoubtedly worshipped with all the zeal and

self-satisfaction of benighted faith. And even now, after
all the proofs furnished at International Exhibitions,
and by every chemist, oinologist, and connoisseur of
repute, there linger, doubtless, in obscure corners of
the kingdom, where the daylight of scientific literature
penetrates not, and where the wine schoolmaster has
not been abroad, comfortable but delusive ideas of the
old-fashioned sort; and it may be, even yet, there exists
in such places a simple and implicit confidence in the
merits of old Port and Sherry Mumbo-Jumbo. Mean-
while, the general public taste has so manifestly altered
that the wine trade is being revolutionised. The strong
old Sherries and Ports of the past are gradually being
supplanted by lighter qualities, which our fathers would
scarcely have recognised as wines. Instead of strong
draughts derived from added alcohol, and cloying sweetness from added saccharum, persons are looking for
wine flavour, and bouquet and cleanness upon the palate.
Much, however, yet remains to be done in order fully to
bring home to the multitude the results of public and
scientific investigations. I think it my duty, as well as
I am able, to collect and arrange a few of the more
important of these particulars. In the first place, I
would point to the International Exhibitions of Wine at
Vienna in 1873, and in London in 1874. Of the latter,
the official report, after stating that " the extent and
collection of wines, especially in regard to the samples
from Spain and Portugal, made it possess an extra
interest and value—giving an insight into the characters
and strengths of the produce of those two countries
never before obtained," furnishes the following table
of the average strengths of natural wines :—

		Lowest Strengths exhibited per cent. of Proof Spirit.		Highest Strengths exhibited per cent. of Proof Spirit.		Average per cent. of Proof Spirit.
Australia	natural	21·4	...	30·3	...	26·39
	fortified	29·2	...	40·8	...	
California		...	...	...	...	24·83
France	natural	16·3	...	27·1	...	20·47
	fortified		...	31·10	...	
Germany	natural	10·9	...	28·2	.	19·41
Greece	natural	17·2	...	28·4	...	23·31
Italy		...	...	...	...	19·69
Portugal	natural	11·7	...	31·2	...	24·07
	fortified	27·0	...	50·	...	
Russia		...	...	...	...	23·43
Servia		...	, ...	...	...	24.27
Spain	natural	10·1	...	30·0	··	24·18
	fortified	21·6	...	56·7	...	

The Commissioners of H.M.'s Customs reasonably enough say of the above figures, that they consider them as " altogether confirmatory of the justice and expediency of the principles adopted, and of the rates fixed at the final settlement of the wine duties in 1862, viz.,—1st, That natural wines might be admitted at an exceptionally low duty; 2ndly, That 26 degrees of proof spirit represent fairly the full strength of almost the whole of the natural wines of the world. They add, that " it is misleading to give the average strength of the fortified wines, because strength produced by the addition of distilled spirit may be carried up to any degree which the interest of the merchant may require, and has actually, in one instance of wine supplied by Spain, been brought up to 56·7—a strength not much inferior to that of old Cognac brandy."

My readers will study the figures of the table with an interest not merely fiscal; they will note that so far from its being the fact that Port and Sherry are so very superior in strength to all the other wines of the world, they are indeed naturally of but ordinary strength, and that there is scarcely any difference between the strength of the strongest Spanish or Portuguese wines, and the strongest Greek or Australian wines, when the Sherry or the Port are imported pure. But then the difference between the pure and the sophisticated Port or Sherry is very considerable, as is indicated by the above-quoted figures. The pure it is difficult, nay, *impossible to successfully imitate;* whereas the Ports and Sherries of commerce, being themselves mixtures, are peculiarly adapted for fraudulent imitation. It is surely worth while to note how much strength Ports and Sherries possess naturally, and how the poorer natural wine always has to receive the larger amount of spirit to bring it up to the strength required by the English merchant. The following table shows the quantity of spirit required to raise the natural to what we may call the commercial strength—viz., 40 per cent. proof spirit, being the usual standard rate of the imported Port and Sherry :—

Natural Strength.	Proof Spirit per cent. required.	Natural Strength.	Proof Spirit per cent. required.
12·0 degrees.	46·4 gallons.	22·0 degrees.	30 0 gallons
14·0 ,,	43·2 ,,	24·0 ,,	26·7 ,,
16·0 ,,	40·0 ,,	26·0 ,,	23·2 ,,
18·0 ,,	36·7 ,,	28·0 ,,	20·0 ,,
20·0 ,,	33·3 ,,	30·0 ,,	16·7 ,,

Now few persons would like it to be said that they drink adulterated wine, nor would any wish to be considered spirit-drinkers, yet, doubtless unconsciously, many are in the habit of taking, under the guise of wine, far more spirit than they are cognizant of. As a rule, spirits are sold by wine merchants from 10 to 17 degrees per cent. under proof, (varying in strength according to age)=to, say 85 of proof spirit. Now as in making grog, persons add from three to four parts of water, it follows that the strength of grog would range from 17 to 21 per cent. of proof spirit, as against 38 to 40 degrees in the fortified wines. But as spirits are not sweetened to anything approaching the extent customary in Ports and Sherries, the grog appears to the taste much stronger than it actually is, as the pungent or acrid nature of the spirit is not then masked by sweetening matter. It should also be borne in mind that the mere mixture of spirit and sugar does not constitute wine; that spirit has no body; and that sugar in conjunction with unfermented wine produces acidity of the stomach.

SHERRY.

BEARING in mind the fact already stated that up to a recent date the produce of Spain was esteemed in this country as altogether far superior to that of France, it is amusing to note the statements of recent advocates of Sherry, in the *Times*, to the effect "that Sherry, when pure, resembles Chablis, Sauternes, Hocks, &c.," and that to the addition of spirit, sweet wine, or sugar, its character is due. Sherry is described in the same correspondence in language which will doubtless be news to many sherry-drinkers as a "liqueur-wine," *i.e.*, "a wine containing a larger or smaller quantity of unfermented saccharine;" and that "the only way to keep this saccharine in subjection is by the addition of extra alcohol." Now, the fact is as certain as anything can be demonstrated, that all these adulterations are absolutely injurious and unjustifiable upon scientific grounds. Only pure wine is wholesome. A well-known authority writes :—

"It is generally agreed, that a wine which is pure and well made is also a wholesome wine, and that an ill-made wine is unwholesome. By a well-made wine we understand one in which all the sugar has been converted into alcohol, and all the nitrogenous ferment has been exhausted and got rid of. If these conditions be present, there is no fear but that the wine will keep, and it will agree with the human stomach. It may not be a strong wine, nor a fine wine, but yet it will keep and be wholesome. On the other hand, an ill-made wine will not keep unless fortified by the addition of a quantity of alcohol sufficient to arrest the further fermentative process. That such wines so fortified are unwholesome is notorious, and modern research has shown that one disease in particular—the gout—is not to be ascribed to alcoholic drinks simply as such, but to such alcoholic drinks as are imperfectly decomposed, and retain in their composition undecomposed sugar and unexhausted ferment."

One of our chief wine chemists, Dr. Thudichum, thus described in the *Times*—

How Sherry is made.

"Sherry—that is to say, the wine grown and made at Jerez for consumption in England—is the product of two varieties of wines mainly, the palomino and mantuo castellano. Each quantity of collected grapes sufficient to yield a butt of must, previously to being trodden and pressed, is invariably dusted over with from 30lb. to 40lb. of burnt plaster of Paris, (sulphate of lime.) The effect of this practice, of which my inquiries among sherry makers have not taught me the object, is to precipitate all tartaric and malic acid of the must, and substitute in their place sulphuric acid. The must, therefore, as it runs from the press, contains no bi-tartrate of potash, or so-called tartar, but sulphate of potash instead. In consequence all sherry contains nearly the whole of the potash of the must as sulphate, amounting to from $1\frac{1}{2}$ kilogramme (about 3lb.) to 7 kilogrammes (about 14lb.) per butt of 484 litres, or 108 gallons.

"The common varieties of must are not only plastered, but also impregnated with the fumes by combustion of about five ounces of sulphur per butt, which adds about a pound of sulphuric acid to that brought in by the plaster. The plastered must as it runs from the press contains its fruit, sugar, tannin, and other ingredients in a perfectly developed condition, and the statement of one of your correspondents that they were in an undeveloped state is scarcely intelligible. Quantative determinations made upon *many and different specimens of must* at Jerez show that its specific gravity varies between 9 deg. and 14 deg. of Baume's areometer, indicating from 14·6 to 24 per cent. of sugar, and that, therefore, it can by fermentation *only form from 14 to 23 per cent. of proof spirit.*"

"The must ferments in the sheds called bodegas, there being no cellars, properly so called, at Jerez. In a fortnight the sugar has all fermented away, and the must is **transformed** into wine.

This is allowed to deposit its lees during some months, and is racked in the following February and March. On this occasion some *brandy is added to the wine*, by which its alcoholicity rises to *about* 29 *per cent. of proof spirit*. In spring and early summer, the wine (still termed "mosto," and so to the time of the next harvest) undergoes what is termed its-first evolution, and after that is ready for further preparation.

"This consists in the addition of various ingredients which impart colour, sweetness, spirit, and flavour. Colour is imparted by the addition of caramel, produced by the boiling down in coppers of previously plastered grape-juice; the brown syrup is dissolved in wine and spirit, so as to form a deep brown liquid, containing from 35 to 50 per cent. of proof spirit, termed 'colour,' or 'vino de color.' Frequently caramel made from cane-sugar is used instead of that made from grapes. Some colour is made with the juice of rotten or otherwise inferior grapes. Sweetness is imparted by 'the addition of "dulce,"—that is, must, frequently made from grapes dried for some days in the sun, to which one-sixth of its volume of spirit, of the strength of 40 degrees by Cartier's alcoholometer, has been added, (a process by which all fermentation becomes impossible.) Every hundred litres of dulce contains, therefore, 19 litres of absolute alcohol, equal to 33·78 per cent. of proof spirit. Flavour is imparted by the addition of some old selected wine, which is kept in so-called 'soleras.' Ultimately brandy is added to the mixture to the extent of fortifying it up to 35 as the *minimum*, most frequently up to 40 or 42, and sometimes, as your Custom-house correspondent proved, up to 50 *per cent. of proof spirit.*

"In a butt of ordinary sherry (40 jars,) there are mostly one-fifth of its volume of dulce (eight jars); consequently, about one-sixth of unfermented grape juice, and which remains unfermented. This is, therefore, opposed to the statement of one of your correspondents, that it would be impossible to find a single drop of unfermented grape-juice in sherry. The better sherries are made less sweet, and only the few finest varieties are left unsweetened. The 'dulce' is never plastered, and therefore its addition depresses a little the large quantity of sulphate of potash introduced by the 'colour.'

" Now it must be observed, that what has been described is the process of making ' sherry,' and not a process of adulterating it. It may be a question whether this process leaves much room for adulteration, or whether it is not itself adulteration ; in other words, whether all sherry whatsoever is, or is not adulterated. To help your readers towards a solution, I remind them that medical authorities have long since pronounced the brandied and plastered sherries to be unwholesome. But the vendors of such sherries are not troubled by the administrators of the Acts of Parliament relating to adulteration. On the other hand, bakers who mix a little alum with their flour or dough, which in the bread re-appears as sulphate of potash (the same as in sherry) and phosphate of alumina (perfectly innocuous), are prosecuted, fined, and denounced, though their additions considered as per cent. of bread are incomparably smaller than the additions made to sherries considered as per cent. of wine.

" Sherries contain from $1\frac{1}{2}$ to 8 grammes of sulphuric acid as potash salt per litre, and the more the older and better they are ; most ' soleras ' are near the higher figure, Now if alum bread is unwholesome, plastered sherry must be unwholesome also, and is more so ; but if plastered sherry be left unmolested by the public health analysts, then alumed bread ought in fairness also to be left unmolested ; it is simply illogical and unjust to punish the baker and let the vintner escape for essentially identical acts.

The same gentleman, in a lecture which he delivered at the Society of Arts, " On Wines : their Origin, Nature, Analysis, and Uses," furnished the following direct statements as to the effect of plastering wines, which it might be well to bear in mind, when we are invited to drink a Sherry whose *spécialité* is that it is free from acid.

The Plastering of Wines Denounced.

" Spanish, Portuguese, and French wines of the south are plastered ; that is to say, plaster of Paris is dusted over the grapes immediately after they are gathered, or while they are on the press, or while they are in the state of must. Dr. Dupré and myself have been unable to find out the logic of that practice. If it is intended to make the wine stronger, it fails; for plaster unites with little more than one-fourth its weight of water; but the gypsum formed encloses mechanically a quantity of must, and reduces the total yield, so that 50 per cent. of plaster will retain fully half the juice, and raise the sugar in the remaining half from 13 to 15 per cent. only, and lesser quantities in proportion. But plaster will diminish the free acid of the wine, in proportion to its quantity, from 5 to 0·5 per mille. It will do more; it will decompose the tartrates, and form sulphates, and thus change wines into drugs. In fact, all Sherries contain considerable quantities of sulphate of potassium, to which many varieties owe their bitter taste, and their purgative action. I am quite open to instruction on the use of plastering, but have sought it in vain of some large producers or importers of Sherry. No doubt, the 20 per cent. of alcohol in Sherry is a frequent cause of kidney affection; but the cause is at least doubled by the potassium salt. I vote for Sherry without plaster acid, and less than 16 per cent. of alcohol; such Sherry will require neither camomile nor nitric ether for a flavour. I vote for not changing ripe must into unripe by removing wine acid, and leaving sour apple acid. I delight in a glass of Amontillado, or even cheap " Vino de Arenas;" but I gladly leave the drink of tincture of Glauber's salts to the old gentlemen who, as the phrase goes, " cannot get anything dry enough."

Dr. Dupré, Lecturer and Professor of Chemistry at the Westminster Hospital, in the course of the discussion which followed upon the remarks of Dr. Thudichum, at the Society of Arts, above referred to, made these important observations on—

The Best Properties of Wine destroyed by "Plastering."

" Moreover, the acids in wine varied considerably. Some contained chiefly tartaric acid,—in fact it was the general superstition that this was the prevailing acid of wine; but this was by no means the case, for Port and Sherry contained scarcely any. Port being of too great alcoholic strength, the alcohol precipitated the acid in the form of tartrate; and Sherry, because of the plastering to which it was subjected, removed nearly all the tartaric acid, and replaced it by sulphate of potassa, a very active saline agent which, like most salts of potassa, had a very depressing action on the heart. Now wine was very frequently given to keep up the action of the heart, which as all physiologists knew was often of extreme importance, and could be effected no way so well as by the administration of alcohol or wine; but it might often happen in the case of Sherry, the slight stimulating action produced by the alcohol would be entirely counteracted by the contrary effect produced by the sulphate of potassa."

The presence of sulphuric acid in Sherry, its general character, and injurious effects, are still further confirmed by the analyses of JOHN POSTGATE, Esq., F.R.C.S., from whom I had the honour to receive the following letter :—

"59, BRISTOL ROAD, BIRMINGHAM,
"*Dec.* 19, 1873.

"DEAR SIR,—I am much obliged to you for your letter of the 13th instant, and the copies of your interesting works on *Wines and its Adulteration*, &c., which I have just received on my return home from the North. I have not published a book on wine, but my analysis of Sherry has been noticed by the Press generally, during the effort I made to procure legislation against the adulteration of food and drugs. My attention was drawn to the adulteration of Sherry many years ago by its effects on the system, producing dyspepsia and pain in the head. I traced these to sulphuric acid and rank spirit in the wines complained of, and analyzed by me in consequence. I exposed the adulteration in my public addresses, and pointed out the results of the use of such impure wine. The Sherry which I was called upon to examine and analyze, from its effects, ranged from 42s. to 72s. per dozen. It all contained free sulphuric acid and fusel oil ; 100 grains of the sample of Sherry at 72s, was so surcharged with sulphuric acid, that when evaporated, the organic matter in it was charred. It was an irritant wine, producing pain in the stomach. Sherry wine is a mixture, and ought to be sold as such, so that pure wine may be known and appreciated : at present, the public taste is adulterated, and, unless the wine is very acid or rank, it fails to detect the impurity.

" I remain, yours faithfully,

"JOHN POSTGATE, f.r.c.s.,

" *Examiner and Professor of Medical Jurisprudence,*

" *Queen's College, Birmingham.*

"*Author of Adulteration Acts*, 1860, & 1872.

Dr. BARRY, in his work on Wines, 1775, thus writes, respecting the—

Medicinal Value of the Natural Tartrates.

" Tartar is the real *essential salt* of wine, which all recent wines contain, but in a very different proportion to the other principles of wine, in which it long remains latent, until, by being more attenuated and disengaged, it then is *successively* separated from the very *centre* of the wine, and equally directed to the *circumferences* of the cask, in which it is kept. This is a real *crystallization* of these natural salts of wine, and, similar to other operations of the same kind, is never formed but when the liquor is kept in a state of *rest*. A much greater quantity of it is separated from the *strong acid* and *austere* wines than from the *rich* sweet wines, in which the fine *oily* and spirituous parts prevail more than the *saline*. On this account the former, while recent, retain a disagreeable austere taste, until the superfluous saline parts are thrown off, and those which are retained become more mild. The rich wines will likewise continue to be disagreeable, luscious, and heavy until the oily and spirituous parts are more *refined* and exalted, to which they entirely owe their peculiar *colour* and *fragrancy*. On this account the wines of this kind, which contain a greater proportion of the saline *refined tartar* than others, acquire a more grateful *pungency*, which eminently improves and distinguishes them, and on that account are called by the Italians the *dolce Piccanti*. Many distempers, and particularly concretions in the joints and urinary passages, are, by some eminent writers injudiciously imputed to this tartar in wines; but this separation of it is a very gradual and slow process, and never can prevail but in a quiescent state, and not possibly while the wine in a quick motion circulates through the body, or in passing through any of the excretory canals. Neither are these calculous *concretions* which are formed in the joints or urinary passages of the same kind with this vinous salt; but really of a different and opposite nature, as it evidently appears from experiments that these calculous *concretions* are of an *alkaline⁻* nature, and this vinous tartar of a penetrative acid kind. The *crystals of tartar*, which are thence formed, are likewise found to be not only a safe, but a useful aperient and attenuating medicine in many cases, and much more apt to attenuate and dissolve such beginning concretions than to form them."

The Vexed Question, Wine Acidity.

One of the greatest difficulties in connection with the improvement of the general judgment in respect of wine, arises out of the inadequate acquaintance of the public with the facts mentioned as above by Dr. Barry. Many members of the medical profession, even, have been misled into giving the preference to fortified over pure wines. In all Sherries, it is now proved, the acid tartrates have been removed by plaster of Paris, and of late matters have been carried to such an absurdity, that the absence of the natural acids in Sherry has been trumpeted far and near as the special recommendation of a certain "Sherry" which boasts of its very deformity, and sets the miserable fashion for thoughtless but humble imitators to follow. I will not myself enter the lists against the Spécialité Sherry, for to do so might suggest merely trade differences, but I will quote the following from a highly influential medical periodical called *The Practitioner*, edited by the late Dr. F. E. Anstie—

The Spécialité Sherry.

" The Rev. Sir Edward R. Jodrell, Bart., in his disinterested zeal for the public welfare, has had the above Sherry analysed at his own expense, and has placed the report of the analyst at the disposal of Messrs. Feltoe & Sons, who make use of it as an advertisement. As this advertisement has found its way into several professional journals, we deem it a duty to our readers to point out the real nature of this much vaunted Sherry. Before doing so, however, we must express our regret at seeing Dr. Redwood's respected name attached to the analytical report in question ; firstly, because we deem the use of such reports for mere purposes of trade exceedingly mischievous ; secondly, because the report in question is full of errors, both of fact and interpretation.

" The total acid is given as amounting to 0·54 per cent. (we presume it is meant that the free acid present is equivalent to 0·54 per cent. of tartaric acid), and it is implied that this is about the usual amount of acid found in good samples of Sherry, in addition to which it is alleged to be the *true acid of the grape*, namely, *tartaric acid*. Now, in the first place, this acidity is rather high for a young Sherry, such as this evidently is ; and, in fact ,assuming the acetic acid to have been estimated correctly, the acidity is, for a Sherry, exceptionally high. In the second place, in the grape juice itself, the greater part of the free acid is usually malic acid, the smaller portion only being tartaric acid. In all wines, even when perfectly natural, the proportion of tartaric acid is still further reduced by the precipitation of the tartar, owing to the alcohol produced by fermentation ; whilst, in all Sherries, the tartaric acid is reduced to a minimum, or is even removed entirely, by the plaster of Paris universally employed in Spain in the process of Sherry making. We are thus reduced to this alternative, either a serious error has been committed in the analysis, or, this reputed Sherry is altogether a factitious article. Again, the proportion of ash is given as 0·45 per cent., and it is stated that this is not more than it should be, and that it contains *nothing foreign to the grape*. Now, if we assume this wine to have been plastered, like all Sherries, this is certainly about an average amount of ash, but then it must contain a large proportion of sulphate of potassium, the greater part of the sulphuric acid of which must be derived from the plaster of Paris employed, and is therefore foreign to the grape. If, on the other hand, the wine has not been plastered, the proportion of ash, is about twice as high as it should be in the natural wine, and we are again forced to the conclusion that we are dealing with a factitious article.

"So much for this most elaborate analysis, as the Rev. Baronet terms it, ironically, as we cannot help thinking.

"Now for the actual facts of the case—

"A sample of the spécialité Sherry, procured from Messrs. Feltoe & Sons, 26, Conduit Street, yielded the following results :—

Specific gravity at 60 deg. Fahr. ... ... 985·9 per cent.
Absolute alcohol by volume ... ... 22·45 ,,
Equal to proof spirit ... ... 39·5 ,,
Total free acid (calculated as tartaric acid) ... 0·44 ,,
Containing acetic acid ... ... 0·15 ,,
Do. Tartaric acid ... ... 0 0.3 ,.
Sugar and extract ... ... 2·01 ,,
Ash ... ... 0·445 ,.
Containing sulphate of potassium ... . 0·405 ,,

These results show that the wine in question is probably an ordinary Sherry, tolerably strongly fortified, and of average acidity. Like all Sherries, it contains scarcely a trace of tartaric acid, but very much sulphate of potassium, and has therefore unquestionably been subjected to the usual plastering."

The scientific theory of the value of Tartaric Acid in wine can best be stated after an analysis of the constituents of this very peculiar product. Tartaric Acid in wine exists in two states, i.e., free, or combined with a base of potash. In a wine report of a recent *Lancet* commission the following results are stated :—

"In the Clarets, the fixed acids, all calculated as free tartaric acid, ranged from 8·82 to 11·65 grains per 1,000 grain measures of wine, the mean being 10·48 ; in the Burgundies, from 8·91 to 12·44, the mean being 10·3̣ ; in the Hungarian wines, from 8·47 to 10·75, the mean being 9·75; and in the Greek wines from 9·77 to 11·03, the mean being 10·50. . . . . The same remark applies with still more force to the quantity of *tartrate of potash present in any wine ;* the greater the amount, *the larger, as a rule, the quantity of grapes employed in the mannfacture of the wine.* Taking the results thus obtained, we find that the tartrate of potash in the Clarets varied, per 1,000 grain measures, from 2·767 to 5·21, the mean being 3·625 ; in the Burgundies, from 2·61 to 4·04, the mean being 3·337; in the Hungarian wines, from 2.76 to, in one sample, 6·17, the mean being 3·688, or, excluding this sample, to 3·19 ; and in the two Greek wines, from 5·9 to 6·01. These figures disclose the unexpected result, that the Burgundies contain less tartrate of potash than the Clarets. How this anomaly is to be explained is not apparent ; probably by a difference in 'the composition of the grape, and in part by the practice of using sugar in the manufacture of Burgundy.

" It will not be out of place here to add a few observations on the effects of the *Acidity of Wine.* It seems to us that there is a good deal of error and prejudice afloat on the subject of the acidity of wine. Acid in wine is regarded by the wine-merchant and by the consumer as something injurious, and even pernicious. We altogether deny that this view of the subject is correct. The acetic and tartaric are both very wholesome acids ; and when they are consumed, as they so constantly are, in salads, and in the grapes and other fruits we eat, they are almost invariably regarded as wholesome and healthful. How comes it, then, that even a minute quantity of these, and especially of the former acid, is held to be so pernicious? We believe that in the majority of cases it is an error thus to view it, and that the acids in good sound Clarets contribute to the wholesomeness of these wines."

Apropos of this question of the Acidity in Wines, Dr. Druitt, in his *Report on Cheap Wines,* published by H. Renshaw, Strand, thus writes :—

C

" Those things are called acid which redden litmus paper, or which neutralise an alkali, or which give a certain impression to the tongue, known as sour. Acids may be inorganic or organic. Amongst the former, the sulphuric, hydrochloric, nitric, and phosphoric are articles of diet or medicine; amongst the latter, the citric, from lemon; tartaric, from grapes, oxalic, from sorrel; the acetic, a product of sugar; the malic, racemic, &c., which exists in fruits; the tannic, or astringent; and the lactic, in sour milk.

"Acids of the wholesome kinds, above-mentioned, are greedily sought for by many persons, and avoided by others, The persons who seek them are usually the young, strong, active, and hearty, with free, open pores of the skin, and good appetites. Acids do to the palate and stomach what soap and towels do to the skin, *i.e.*, they strip off its coating, make it redder, more active, and ready to secrete. Hence the love for lemon-juice, vinegar, and pickles at dinner, and the charm of acids to persons in certain kinds of bad health, torpid liver, coated tongue, &c. The secretions of sore throats are alkaline, and an acid liquor wipes this off, and leaves the surface clean. The persons who avoid acids are usually the torpid, and those with red tongues, or skins locked up. In good wine the acidity is due to tartaric and volatile acids, each wholesome *per se*. If too acid, the fault may be excess *simpliciter*, or, more probably, defect of *body*, which should veil the acid. The only test of *quantity* of acid is the chemical one; and this shows that very first-class wines of the Rhine and Moselle contain most acid—Port and Sherry, least. But it must be remembered that one-fifth, or more, of Port and Sherry is not wine, but spirit; and, secondly, that the makers of sham wine can put in as little as they like, or can neutralise natural acidity by chalk. Hence, quantity of acid is no test for quality of wine."

These opinions are confirmed by the late author of the work *On Foods, &c.*, Edward Smith, M.D., F.R.S., in his letter to the *Times* on " The Alleged Adulteration of Sherry," who thus writes :—

" The removal of the natural salts of grape-juice—viz., the

tartrates and malates—by the plastering process, greatly deteriorates the wine in one of its nutritive, or, to speak more correctly, medicinal qualities, while the formation of sulphuric and acetic acids increases the evil. There are qualities in wine made simply from the juice of the grape, which are not destroyed by the quantity of alcohol generated therein; and changes proceed in such natural wine of a valuable medicinal and dietetic character much more rapidly than in adulterated Sherry or changed wine. These are well exemplified by a natural white wine of Portugal—Bucellas—and the manufactured wine of Spain—Sherry. Nothing is more desirable in this discussion than to satisfy wine consumers that while pure wine must contain alcohol. Alcohol is not wine, neither is it the most important element of wine in a dietetic sense. A sound Claret is far better for health than a sound Port, and cheap white Greek wines, than dearer Sherry."

From the foregoing it will be manifest that the tartaric acid in wine greatly adds to its nutritive and dietetic value; consequently, when wine is described and sold as *without acidity*, it is deficient in one of its most valuable properties. However unwelcome the fact may be, the goodness, fine bouquet, and flavour of every wine are derived from and produced by the unhappy bugbear, acidity; that is to say, by the oxidation of the alcohol certain *ethers* are developed, which are the cause of the value and bouquet of old wine; for we all know that in judging, smelling, and tasting an old-bottled wine, we value it neither according to its alcoholic strength, nor sweetness, but to its vinous flavour and bouquet. Thus we find it stated by Mr. Griffin, in his *Chemical Testing of Wines and Spirits,* that—

"When it is distilled the acetic ether passes over with the alcohol into the distillate, where it can be detected by its odour, but cannot be measured as to its quantity. Thus it is that as wines grow old, these ethers are produced in them; and to these ethers is no doubt attributable all the odour of the wines, and much of their flavour. Such is the theory of bouquet, which appears to me to be most consistent with our present knowledge of the constitution of wines. . . . . . Although, within certain limits, wines are powerful and good in proportion to the quantity of alcohol they contain, yet, when taken alone, the quantity of alcohol affords no indication of the value of the wine. It is *one element* of goodness, but not sufficient in itself to constitute goodness. As a decisive proof of this fact, I may refer to the wine No. 12, in Tables I. and II. This contained 50 per cent. more alcohol than Fresenius's best Steinberger, but it was detestable rubbish, bought of a London grocer for a shilling a bottle, and not worth a penny. Hence, predominant alcohol, considered *alone*, is no mark of goodness in a wine. . . . : It has been asserted by Professor Fresenius that the goodness of wine is so much the greater the more it contains of sugar. In this he was deceived by the circumstance that his wines contained much saccharine, because they were prepared from very ripe grapes, and had not been completely fermented. Good wines of the light sort—and those of the Rhine belong to that category—should have, and indeed usually have, either no sugar, or very little. All the sugar that occurs in the grape-juice ought to be entirely fermented into alcohol; and if this is done properly, and no injurious substances are introduced into the wine, and the wine is so well fermented as to have the alcohol in proper proportion to the acid, it will be sweet enough for agreeable drink without containing the least particle of free sugar. If sugar is practically left in wine, you may take it as a proof of incompleteness of fermentation, or that the manufacturer meant to produce a sweet wine, or that he was conscious of the presence of some improper substance or quality in the wine, which he considered it necessary to cover or hide by sweetness—*mere sweetness*, the lowest, and most easily and cheaply imitated, of all the qualities by which wine is characterised; consequently, the presence of abundant sugar in wine is no proof of superior goodness. These reasons show that the goodness of wine is

neither caused by a small proportion of acid, nor by a large
proportion of sugar, and, we have previously seen that it is
not decidedly influenced by a preponderance of alcohol. . . .
The true principle which, more than all others, appears to
regulate and govern the goodness of wines is, that *the weight of
their alcohol should have a certain relation to that of their acid.*
If this relation is right, other things are comparatively
indifferent."

But besides the tartaric acids, or tartrates, above
referred to, there is another important acid in wine,
namely, the tannic acid, or tannin, which is the astringent, or rough principle that is extracted from the
skins, pips, and stalks of the grape, during the process
of fermentation, and is found chiefly in red wines.
Tannin not only adds greatly to the keeping properties
of all wines, and is productive of flavour, but is highly
important, in a dietetic point of view, for its tonic and
strengthening effects. The medical profession has long
recognised the fact that the wines best fitted to aid a
languishing digestion are those that are rich in tannin,
and contain the greatest amount of alcohol—not alcohol
added, but that which is produced naturally in the
course of fermentation. Thus the learned chemist of
Strasburg, Dr. Schelagdenhauffen, states that,—

"Peruvian bark in wine is beneficial in several forms of
anæmia, &c.; but experience has proved that not more than
one-fourth of the alkaloids of the bark are dissolved, and that
which the wine dissolves is the tannin. The wine most rich in
natural tannin should therefore be preferred to that prepared
with Peruvian bark. . . . In all these conditions a small
glass of rough wine—*i.e*, containing tannin—at the end of a
meal is the best cordial to animate the energy of the digestive
functions, and re-establish harmony in the great organ of
nutrition."

The next important consideration. is the amount o

alcohol, or spirit, that grapes are capable of producing,
and the cause that influences its quantity.   Dr. Faure,
in his *Analyse des Vins*, remarks that—

"Alcohol being the product of the decomposition of sugar
during the process of fermentation, it is evident that the sweeter
the grapes the more alcohol they will produce, and the resulting wine will be strong and generous.   It is not only on
account of the quantity of alcohol produced that ripeness of
the grape is indispensable for making good wine, but because
the chief sugar being the result of the last change which takes
place in the fruition, the other constituents have also been
developed *pari passu*, so that in perfectly ripe grapes all the
elements of good wine will be united.   The contrary will
occur should the season prove wet and cold, for, in that case,
the wines will be weak, colourless, and without bouquet and
flavour.   Thus, then, although alcohol may be one of the most
essential elements in wine, one must not suppose that *it* alone
constitutes its quality, or that it is capable of producing the
beneficially mental and physical effects derivable from the use
of pure wine; were it so, plain diluted spirit would answer
every requirement."

To the same effect, we hear from Dr. Druitt, that—

"Alcohol has plenty of sins to answer for.   It produces
dropsy, delirium tremens, disease of the brain, liver, and kid-
neys.   But to produce gout, it is not alcohol *per se*, but in
combination with sugar and ferment, that is needed.   Of such
combinations, sweet ale, sweet cider, and sweet wine, are well-
known examples, provided their sweetness be the result of
arrested fermentation.   It is the interest of the physician, then,
no less than of the politician, we will not say to *discourage*, but
certainly not to *encourage*, the production of imperfectly fermented liquors, by giving them artificial protection in the
shape of exemption from their fair share of customs or excise
duty.   The most popular example of an ill-made and unwholesome wine is modern Port.   In fact, it does not deserve the
name of wine, but rather of *liqueur*, and so the French call it."
—*Medical Times and Gazette*, Oct., 1867.

From Dr. Faure's statements it will be seen that the alcoholic strength of *natural wines* depends upon the ripeness and sweetness of the grapes ; and that without perfect fermentation even the best grapes will not produce much alcohol, rarely exceeding perhaps, 26 to 28 per cent. of proof spirit.   It may be noted that, according to the various latitudes, the wines of France vary in strength from 9 to 24 per cent. of proof spirit ; those of Spain and Portugal, from 11 to 28 per cent ; those of Germany and Hungary, from 9 to 21 per cent. ; whilst those produced in Greece, aided by a favourable clime and limited area, are more uniform than those of any other country, varying but from 23 per cent. to 29 per cent. of proof spirit.

Dr. A. H. Hassall has not only stated the following as the results of his own experiments with Sherry, but has actually patented a process, hereafter mentioned, for, what he calls, the deplastering, or restoring to Sherry its natural constituents.

" I have subjected, he says, " nineteen samples of Sherry to full quantitative chymical analysis.   Of these nineteen samples eight were of the highest quality procurable, and their analysis was undertaken with a view to arrive at certain standards by which the other samples, purchased in the ordinary way from wine merchants, restaurant proprietors, and publicans, might be compared.   The results arrived at were as follow :—

" 1. That the whole of the wines, without exception, were fortified with extraneous spirit to a large extent.   This spirit, doubtless, in nearly all cases, and probably in every case, is derived either from corn, beet-root, or potato, and not from the grape ; while the average amount of proof spirit furnished by the must, from which Sherries are made at Xeres, according to the best authorities, is about 19 per cent., the lowest quantity found by me was 29·723, and the highest 41 294, the mean of all being 35·477 per cent.   In fact, the quantity of spirit added falls not very far short, of that actually furnished by the fermentation of the grape-juice itself.

"That seventeen of the nineteen samples were decidedly plastered.   The quantity of sulphate of potash found in the wines, after deducting three grains per bottle—this being the utmost amount ever met with in natural Sherry—ranged from 15·0 to 51·6 grains per bottle.   These quantities give 90·0 grains as the lowest, and 309·6 grains as the highest amount per gallon. It will be seen, therefore, that these analyses bear out the statement of Dr. Thudichum,—that all the Sherries imported into this country are plastered ; that is to say, the must is dusted over with sulphate of lime ; in addition to which, it is also impregnated with the fumes of burning sulphur, whereby a still further quantity of sulphuric acid is introduced into the wine.   Dr. Thudichum gives the quantity of sulphate of potash contained in Sherries as varying from 36·1 to 169·2 grains per bottle of one-sixth of a gallon.   It will be seen that my highest quantity amounts to 51·6 per bottle, or 309·6 grains per gallon, equal to about three-quarters of an ounce : the quantity of sulphate of potash, therefore, met with in these analyses is much below the larger amount given by Dr. Thudichum— namely, nearly $2\frac{1}{4}$ ounces.

" 3.   That, in addition to the fortifying and plastering, five of the wines contained considerable amounts of cane sugar, the presence of which affords, of course, clear evidence of adulteration.

"That two of the Sherries—those denominated 'Hambro' Sherries—contained very little wine at all, but consisted chiefly of spirit, sugar, and water, flavoured.   In fact, these mixtures could hardly be said to have any claim to be regarded as wines at all.   It will thus be seen that, notwithstanding that eight of the samples were of the highest quality obtainable in this country, not one of the wines can be regarded as the pure and natural product of the grape alone."

It is, perhaps, a not unusual experience that science of a certain kind is prone to consider cure rather than prevention ; but one scarcely expects to find men of

such high scientific position as Dr. Hassall, after de-
nouncing the plastering process, providing a remedy for
the manufactured disease. The *Medical Times and
Gazette* makes merry over the patent which Dr. Hassall
has taken out for taking the plastering out of
Sherry :—

" Everybody knows that Sherry is not like Bordeaux, a "self-
making" wine, as Messrs. Gilbey call it (a), consisting of the
juice of the grape left to ferment and clear itself, but is more
or less a manufacture. In very many vineyards they begin by
sprinkling the grapes with plaster of Paris; then the grapes
are trod, the juice, expressed, fermented wholly or in part,
mixed with a certain quantity of spirit, put by to mature,
sometimes sweetened with boiled grape-juice, and stored in
huge butts, where reserves are kept for scores of years, improving in flavour: then when the time for shipment comes,
various proportions of the newer wine, with a share of the
older wine, and of sweeter or drier, and perhaps more spirit,
are blended together to suit the various tastes and pockets of
the consumers.

"Sherry may be dark as mahogany, like the old brown Sherry
of King George IV.; or, may be, of the palest amber; it may
be light and dry, or rich and sweet; but, in any case, good
Sherry is soft, has no taste of raw spirit; is cordial, yet not
hot, and has an agreeable fragrance, and the indescribable
heart-and-brain-sustaining, and soul-and-body-comforting flavour of old wine. Such wine is emphatically good and wholesome in the proper cases, no matter whether it is natural or
manufactured, whether plastered, or not plastered—if the product is good, the manufacture cannot be blamed.

" On the other hand, we may have Sherry dry or sweet, pale
or dark, but hot, strong, and spirituous, with no grapy taste.
Perhaps, if pale and dry, it is dosed with nitrous ether, to
imitate the 'amontillado' flavour; if not strong, it is probably
flat and flavourless.

" If a man chooses to drink bitter, flat Sherry, perhaps he
likes sulphate of potass! There is no accounting for tastes.

This hubbub about sulphate of potass in Sherry is grotesque enough, but the most ridiculous part remains. Whether some Samaritan friend wished to take pity on the poor plastered wine, or whether some enemy of the Sherry trade wished to damn it by publishing an elaborate process for restoring the wine to a natural state, seemed uncertain from a first perusal of the following advertisement:—

"'WINE.—Deplastering and Improvement.—Sherry is always subjected to the operation of plastering, whereby the wholesome tartaric acid is removed, and the aperient sulphate of potass substituted. The advertisers have secured a patent for the restoration of the original tartaric acid, and the removal of the sulphate of potass by a simple process, and are prepared to treat for the sale of the PATENT or for royalties. Wine thus treated is greatly improved.—Address, Medicus, St. Catherine's House, Ventnor.'

"This advertisement must have been startling enough to those who put their faith in Sherry. But it is nothing to the statements in the "provisional specification" of the patent No. 2706, which has been taken out by Arthur Hill Hassall, M.D., and Otto Hehner, analytical chemist. In this document it is declared, amongst other things—1, that "many wines contain as much as 500 grains, or over one ounce, of sulphate of potass per gallon;" 2, the process for deplastering and improvement of wine is declared to be the addition of tartrate of baryta to the wine."

After describing the method and estimating the effects of Sherry before and after such a proposed process, the Editor of the *Medical Times and Gazette*, says—

" That this process may answer in the delicate hands of analytical chemists, if they condescend to dirty their fingers with poisonous Sherry, is one thing; that it should ever be largely adopted in the trade, is another. With new chemical processes, new forms of poisoning arise. All soluble salts of baryta are poisonous; and who shall say that the tartrate of baryta might not be used in excess by unskilled hands, and some of it be dissolved ? What a deplorable picture is drawn of plastered Sherry if it needs this resurrectionising process !"

" As we said before, Sherry is a liqueur or cordial, rather than a beverage. If used, it should be good. Any solution of sulphate of potass is bitter. No good wine tastes bitter. Instead of a doubtful light Sherry plastered or de-plastered, it is better to drink a natural wine with all its tartaric acid, such as St. Elie, or some white French or Hungarian wine."

The *Sanitary Record* is still more adverse to the principle of the patent. It asks—

" Are the public protected against ingenious inventions for improving the quality of food by chemical processes ? We think not, or we should not now have to direct attention to that unfortunate liquid, which is known to the British public under the name of ' sherry.'

" It is well known that the Spanish wine-producers have a practice of adding plaster of Paris either to the grapes or the juice expressed from them (the must). The object of this addition of a mineral ingredient, like sulphate of lime is not very apparent, while its disadvantages are manifest. It tends to remove the natural acid of the grape, namely the tartaric, and to substitute for it a mineral acid, the sulphuric, forming with the tartar, the sulphate of potash which remains in the wine after fermentation. The process of sulphuring for arresting or reducing fermentation, also adds a quantity of free sulphuric acid to the wine.

" We have now before us a copy of a provisional specification for improving the quality of wine, especially sherry. From a description of this purifying process, we find that the object of the invention is to remove from Sherry the sulphuric acid and the sulphate of potash, as well as any free sulphuric acid in the wine, by the addition of the tartrate of barium. The tartaric acid removed by the plastering process is said to be thereby restored to the wine, while the sulphuric acid is precipitated under the form of sulphate of barium.

" By a strange misnomer the patent process has been called the ' de-plastering ' of Sherry.; and wine thus treated has been described as absolutely free from plaster ! It is said to be soft in flavour and to have a much higher money value than the plastered wine.

-" The invention is ingenious, but it appears to us the ingenuity is misplaced. The proposition to remove a non-poisonous salt by a poisonous barium-compound, is certainly a novelty in the treatment of Sherry. It is hard to say what amount of doctoring this wine has already undergone in Spain and England before it reaches the consumers. The plasteriug, sulphuring, flavouring, colouring, and fortifying with alcohol, are accepted stages of manufacture to which the discontented Briton must resign himself. To add to these numerous processes another for ' de-plastering,' as it ts termed, by the addition of a poisonous salt, is carrying matters beyond all reasonable bounds.

" Nothing, in our opinion, will justify the alleged purification of an article of almost universal consumption, by the use of a pernicious compound like that which forms the basis of this specification. A wine merchant, who had the candour to inform his customers that his Sherry had undergone the ' barium process ' of purification, but that the greatest care had been taken to prevent any of the noxious substance remaining in the wine, would hardly reconcile them to the use of it.

" Most persons would prefer the alleged ' aperient' and ' depressant ' action of the sulphate of potash to the possible introduction of any poisonous ingredient.

" The proper correction of the evil would be to prevent the use of plaster at the fountain head, and not to rely upon scientific schemes for removing it or its products after it has been once introduced. Chemically speaking, it cannot be a necessary step in the manufacture of the wine, and it is highly probable that it alters the flavour and quality of the fermented product.

As a matter of course, the imitations of Sherry imported from Hamburgh into this country contain even less wine than the Spanish Sherries, to which the above analysis applies. Dr. Hassall subsequently analysed some Hamburgh, or Elbe Sherry, at the instance of its importers, with the result that the Hamburgh wine was found to consist of sweetened and fortified wine to the following extent :—

"No. 1 in the subjoined table represents the original natural wine, and Nos. 2, 3, 4, samples of the same wine in different stages of manufacture :—

	1.	2.	3.	4.
Specific gravity	0·9973	0·9897	0·9840	0·9884
Absolute alcohol.........	5·555	11·154	25·055	18·231
Proof spirit	11·282	22·652	30·575	37·025
Acetic acid	0·087	0·041	0·009	0·019
Tartaric acid	0·637	0·459	0·484	0·439
Sulphuric acid...........	0·022	0·023	0·022	0·026
Phosphoric acid	0·030	0·029	0·031	0·021
Total solids	1·252	1·508	1·468	6·939
Mineral matter	0·214	0·152	0·418	0·088
Alkalinety, equal to ...	0·052	0·026	0·023	0·033
Glucose	0·294	0·194	0·260	4·527
Cane-sugar	...	...	...	0·430
Nitrogen	0·018	0·010	0·022	0·017

The wine thus analysed must have been of a very poor kind, for it seems to have been fully fermented, there being in it but a trace of glucose. It was, therefore, originally what one would call a dry, and not a sweet wine. By the time it became the " complete product," in column 4, its alcoholicity had been raised to 37 per cent. of proof spirit, to attain which, it required 41 gallons of proof spirit, and, besides that, had received 4·5 per cent. of glucose, or grape sugar. From the above data, it may be accepted that Hamburgh Sherry in its natural state contains 11·282 degrees of proof spirit, and that it is so made that its commercial strength is 37·025. Taking this, the most favourable statement that can be made, of the value of the original wine, and working out the figures as follows, we find—

$40\frac{978}{1000}$ gallons of proof spirit per cent. to be added.

PROOF.

100 gallons, containing .   ...   11·282 per cent. of proof spirit.
$40\frac{878}{100}$ „   of spirit added . . 40·878

$140\frac{878}{100}$ „   of Hambro', contng. 52·160   „   „
As $140\frac{878}{1000}$: 100 :: 52·160 : 37·025

37·025, the strength arrived at after adding 40·878 per cent. As respects the glucose, above-mentioned, it may be as well to add a few words. Dr. C. Graham, of University Hospital, in one of his Cantor Lectures at the Society of Arts, in 1873, said—

"Glucose is now made upon a very large scale by the conversion of starch into grape sugar. . . . . At one of these companies I saw the whole process, from the beginning to the end. I found that they were using rice, ground very small; it was then mashed with water, containing 1 per cent. of oil of vitriol. I believe the other company uses some other form of starch; but it makes very little difference, except that potato-starch, and the cheap inferior starches of that kind, are rather liable to unpleasant oily bodies (fusel oils), and, therefore, they should not be used."

The following extract from the *Times*, of August 15, 1874, shows that the use of glucose is not simply confined to the sweetening of Hambro' Sherry, but is used for making Hock and Moselle :—

"The Cologne Chamber of Commerce, in its yearly report which has just been issued, complains of the adulteration, or rectification, as it is called, of German wines. This, it says, assumed alarming proportions last year among nearly all the vineyard proprietors of the Moselle, and among many makers of the Palatinate. Unsugared natural wines are now scarcely to be met with in the Moselle district, and the addition of sugar goes hand in hand with liberal dilutions of water, and the usual ingredient of spirits. The mixture is formed with grape husks; it is then styled wine. Last season 18,000 centners of common potato sugar were despatched from Coblentz up the Moselle, and considerable quantities were sent to the Upper Rhine; so that many cellars now contain more 1873 wine than the vineyards actually produced."

Of the wines of France it must be remarked that these have acquired, and are acquiring, much of the good opinion and public support in the British markets; which those of Spain and Portugal have been gradually but surely losing—as witness the following plain and incontrovertible figures, supplied by the Board of Trade in their official returns:—

Relative Proportions of the Wines of Portugal, France, and Spain, taken for Home Consumption, 1794 to 1874:—

Period.	Portugal.	France.	Spain.	Period.	Portugal.	France.	Spain.	Period.	Portugal.	France.	Spain.
1794	75·67	3·26	16·67	1845	39·91	6·58	37·93	1865	23·58	21·64	43·11
1796	69·44	0·81	26·11	1850	43·73	5·29	38·36	1866	22·62	25·71	41·49
1800	75·90	0·99	19·43	1854	34·39	8·12	38·34	1867	20·79	26·16	42·73
1805	65·55	1·88	25·44	1858	28·69	8·54	39·67	1868	18·83	29·76	40·98
1815	59·92	4·34	21·50	1859	27·82	9·58	39·60	1869	19·00	26·15	41·13
1820	51·49	3·58	20·46	1860	24·14	15·30	40·44	1870	19·5	27·9	40·12
1825	52·45	6·56	22·86	1861	25·06	20·65	37·38	1871	19·6	27·1	39·11
1830	44·60	4·79	32·35	1862	23·97	19·38	40·35	1872	19·5	28·3	41·1
1835	43·30	4·23	34·74	1863	25·00	18·37	43·29	1873	19·3	30·16	38·14
1840	40·72	5·21	38·16	1864	22·52	20·08	43·47	1874	21·0	28·13	39·9

The foregoing figures are conclusive of the advance which has been made in favour of pure vintages. The only drawback to the satisfaction with which all wine reformers must contemplate such statistics is derived from the knowledge that Bordeaux wine-makers are imitating the examples of Xeres and the Alto-Douro, and mixing with the natural and wholesome wines of the Medoc the coarse, heavy, and loaded liquids of the South of France.

It would be too much to expect that a change so great as that evidenced by the foregoing tables could be made in the wine trade without some endeavour to stem the tide of public opinion ebbing so powerfully in the instances of Port and Sherry. It is, perhaps, worth while to note some of these endeavours as of value from an historical point of view, and I therefore mention two of the most recent and the most influential. Messrs. Ridley, in their well-esteemed Wine Trade Circular of January, 1876, the organ of the Port and Sherry interests, comfort their clients by asserting, despite the steady decline shown by the Board of Trade returns, that Sherry is being " re-habilitated " in public favour. Whether Messrs. Ridley's hope is in Dr. Hassall and the doctor's novel " reviver " cannot be certainly pronounced ; but if so the hope is delusive. People will no more buy re-habilitated Sherry when pure Sherry is to be had, than they will purchase articles " better as new " in the old-clothes line, when they can obtain properly-made habiliments of the latest fashion, requiring no patent " reviver." I do not, however, myself consider the word " rehabilitate " so ill-chosen as some might, for I bethink me of the difficulty of adapting the queer habits of gouty old port-drinkers to modern ideas, and of that still greater difficulty which Port and Sherry "outfitters" have to overcome in the re-habilitation of those "coats of the stomach" which added spirit wears, so to speak, threadbare and seedy. It is, however, more satisfactory to find that Messrs. Ridley do not leave us absolutely ignorant as to the Sartor who is performing the feat of "re-habilitation of Sherry. The holes which have been discovered in Sherry's character have been all made, Messrs. Ridley say, by "mendacious" writers interested in detraction. "A Daniel" has at last "come to judgment" in the person of the respected contributor to the *Pall Mall Gazette;* and all the writers in the *Times,* all the critics and the connoisseurs must bear to be first of all condemned as ignorant and "mendacious," and then it is to be supposed converted to the true faith. Let us turn, then, to the new wine doctrine according to the *Pall Mall Gazette,* and see how far it bears out such assertions. Instead of the plastering process being denied, we find the writer boldly admits it, but pleads "it is not harmful," and says "*if any one imagines that it operates to the extent of changing the flavour of the wine in the smallest degree, or that Sherry, fermented without having had gypsum added to it, possesses any of that fresh acidulous flavour which sound judges so much admire in wines of more northern latitudes, he is greatly mistaken.*"

So, Plaster of Paris, we are taught, does neither harm nor good; nobody can tell the difference. How does that fit with Dr. Hassall's views? If there be no difference we are all mistaken; but will it be said that Sherry is restored to good opinion by such statements

D

as this which is quoted directly from the supposed " re-habilitating " articles referred to by Messrs. Ridley :—

" *My own opinion is that there is something radically wrong in the mode of fermenting wines in the south of Spain.*

" Our wine merchants are mainly responsible for any excess of added spirit to the higher class sherries. Over and over again we were told that they positively demand it of the shipper, who, if left to himself and not made responsible, as he most absurdly is, for the condition of the wine for two years after it leaves his possession, would send it over containing several degrees less of spirit. *It is excess of added spirit, and not gypsum or sulphur, which is the real bane of sherry.*"

PORT.

IT will be worth while after having brought down the Sherry history to its latest chapter, to turn attention to a few facts about Port. As to what Port really consists of, no one is better authorised to state than was the late Baron Forrester, himself a wine-grower of considerable experience. He thus describes the mode in which Port wine is treated before it is despatched to England. How it fares subsequently we shall see hereafter :—

How Port Wine is Made.

" To produce *black*, *strong*, and *sweet* wine, the following are the expedients resorted to : The grapes being flung into the open stone vat indiscriminately on the stalks, sound or unsound, are trodden by men till they are completely mashed, and there left to ferment. When the wine is about half fermented, it is transferred from the vat to tonels, and brandy (several degrees above proof) is thrown in, in the proportion of twelve to twenty-four gallons to the pipe of must, by which the fermentation is greatly checked. About two months afterwards this mixture is coloured thus : A quantity of dried elderberries is put into coarse bags ; these are placed in vats, and a part of the wine to be coloured being thrown over them, they are trodden by men till the whole of the colouring-matter is expressed, when the husks are thrown away. The dye thus formed is applied according to the fancy of the owner,—from twenty-eight to fifty-six pounds of the dried elderberry being used to the pipe of wine. Another addition of brandy, of from four to six gallons per pipe, is made to the mixture, which is then allowed to rest for about two months. At the end of this time it is, if sold, (which it is tolerably sure to be after such

D 2

judicious treatment,) transferred to Opoito, where it is racked two or three times, and, receives two gallons more of brandy per pipe, and then it is considered fit to be shipped to England, it being about nine months old ; and, at the time of shipment, one gallon more brandy is usually added to each pipe. This wine having thus received, at least, twenty-six gallons of brandy per pipe, is considered by the merchant *sufficiently strong*, —an opinion which the writer, at least, is not prepared to dispute."

The Honourable Robert, now Lord Lytton, late Her Majesty's Secretary of Legation at Lisbon, and now Viceroy of India, thus reported on the manufacture of Port wine and the quantity of spirit added thereto :—

" I have frankly submitted to the judgment of Mr. Johnstone, of the Testing Department of the London Custom House, my own estimate of the quantity of adventitious spir admitted into the composition of Port wine, and that gentleman not only assures me that my estimate is a moderate one, but he has also had the kindness to favour me with his own, derived from long observation of the results of the application of the alcoholic test to Port wines, since that test was first adopted to the present day, as well as a thorough knowledge of all the details of the manufacture, and a comprehensive and impartial examination of all existing evidence upon the subject. I subjoin this estimate. ' I find,' says Mr. Johnstone (writing to me in reply to my questions upon this subject), ' that the strength of the spirit commonly used in Portugal varies from 45 per cent. O.P. to 50 per cent. O.P., and I assume it at its lowest, viz., 50 per cent. But the German spirit now so largely imported for fortifying purposes into wine-growing countries, is often as high as 70 per cent., and rarely below 67 per cent.

" ' The composition applies, in this instance, to the higher qualities of Port wine. To the half-fermented wine there is added, to check fermentation, first,—

Gallons of proof spirit.

25 gallons brandy, at 45 degs. equal to 36·25

And say, 5   ,,   geropiga

then,   6   ,,   more of brandy ...   ,,   8·70

again,   2   ,,   more after racking   ,,   2·90

and,   1   ,,   more on shipment .   ,,   1·45

39 liquid gallons   ......   ,,   49·30

76 of wine.

Total 115 gallons of Port wine.

···'That would be,' he continues,' ' of proof spirit, *upon* the pipe of 115 gallons, a little above the maximum of 42⁰, at the higher duty of 2s. 6d.' ''

The above statement having been characterised by Mr. Elles, of Oporto (in the recent correspondence as to the '· Alleged Adulteration of Sherry,'') as a '' gross exaggeration,'' *The Times* inserted the following letter from me (in reply to Mr. Elles' allegation,) enabling me to prove the truth of (the now) Lord Lytton's remarks.

Manufacture of Port.

Mr. James L. Denman writes to us :—

'' As the writer of the letter to which, in *The Times* of the 5th instant, Mr. Elles refers, I shall feel obliged, in justice to myself, if you will allow me space to show that the particular calculation made by (the now) Lord Lytton, as to the amount of spirit added to Port wine was quite correct, and not a 'gross exaggeration,' as asserted by Mr. Elles; also, that his Lordship had allowed for the amount of spirit generated in the half-fermented wine. According to the Parliamentary reports issued in 1862, relative to the natural strengths of wine, Port, when fully fermented, makes but 25·9 per cent. of proof spirit. Lord Lytton gives 13·4 per cent. as the natural strength of half-fermented Port wine, to which is added, per pipe of 115 gallons, to check fermentation, 39 over-proof liquid gallons, equal to 40·3 gallons of proof spirit, as may be thus shown :— 115 gallons half-fermented, containing 13·4 gallons per cent. of proof spirit, equal to 15·4 gallons of proof spirit per 115 gallons; add 39 gallons over-proof liquid, equal to 49·3 gallons of proof spirit ; total, 64·7 gallons of proof spirit. If, therefore, 154 gallons contain 64·7 gallons of proof spirit, what do 100 gallons contain? 42 per cent.; viz., the quantity stated by Lord Lytton. Mr. Elles furthermore says, that Lord Lytton's Reports were clearly controverted at every point by Mr. Crawfurd, Her Majesty's Consul at Oporto :—Considering that, when fully fermented, the highest strength is but 26 per cent. of proof spirit, it sounds rather strange to hear 'that, after a series of very careful inquiries in this city (Oporto), my opinion is that 39 per cent. is about the average proportion of proof spirit to wine. Some few wines, mostly new, may contain as little as 36, and, again, a few richer ones 41, or even 42 and 43 degrees. It must be remembered 1 per cent. is added on shipment. A gentleman from Lisbon, at present employed in this city, is investigating, on behalf of the Portuguese Government, the strength of the wine now held in stock : although his report has not yet been published, it is well known here that his experiments show the average strength to be 40 per cent., or 1 degree higher than I have given it.' It will be seen that the above extract from Mr. Crawfurd's Report confirms Lord Lytton's statement as to the imported and made-up strength of Port wines."

Mr. Vizitelly on the Character and Quality of Port.

[FROM REPORT OF THE VIENNA, 1873, WINE EXHIBITION.]

"Unfortunately, Port, as we know it in England, is, at its best, a dull, heady wine—depth of colour, and a certain fulness and roundness being its principal merits ; for its bouquet, in lieu of the fragrance of fruit or flowers, has too often an odour of ardent spirits, whilst its warmest admirers would never claim for it either raciness or freshness of taste. It is, moreover, especially deficient in *finesse*, is altogether lacking those subtle gradations, and that refined harmony of flavour, that combined freshness and softness which distinguish the grand *crus* of the Haut Medoc; added to which, unlike those unique wines, it· leaves neither the head cool nor the tongue fresh. . . . . One whose knowledge and experience entitle him to speak with authority on the subject of pure *versus* fortified wines, and more especially on the wines of Portugal, writes to me as follows :—' To my mind it is perfectly certain that the fortified Port which is now drunk, will in due time disappear altogether from consumption. . . . . The British wine-merchant and the ignorant consumer are at the bottom of all the mischief. They demand a fruity young Port for bottling, in the fond hope that after twenty years the wine will have become dry in bottle. *Pas du tout.* After twenty years it remains just as sweet, if the shipper has added enough brandy, otherwise it will be *vin aigre.* . . . . Now-a-days spurious Port is produced on a large scale at Tarragona, in Spain, which imports considerable quantities of dried elderberries, presumably for deepening. the colour of, if not for actually adulterating, the so-called "Spanish Reds." A couple of years ago I tasted scores of samples of fictitious Ports in every stage of early and intermediate development, rough, fruity, fiery, rounded, and tawny, in the cellars of some of the largest manufacturers at Cette, and saw some thousands of pipes of converted Rousillon wine lying ready for shipment to England and various northern countries as vintage Port.' "

Why Sweet Wines produce Acidity.

" It has been supposed," said Dr. Lankester, the late Coroner for Middlesex, in his popular *Lectures on Food*,—

"That acid wines are bad where there is acidity of the stomach. But there are two sorts of acid, or, I may say, three. There is tannic acid, which gives the astringency to red wines, and is the principal agent in the formation of the crust ; then there is the tartaric acid, which gives acidity to wine; and there are the acids which, uniting with compounds in the wine, form the flavour and bouquet of wines. It is to the tartaric acid I would now draw your attention. The tartaric acid is the acid which distinguishes the fruit of the grape; it occurs in varying quantity in grapes, but it is always found in wine made from grapes. Now, this acidity of the stomach more frequently arises from the decomposition of sugar than from anything else; and wines which have sugar enough to cover their acidity have been taken to prevent this state of the stomach, whilst acid wines which contain no sugar have been avoided. Neither tartaric, acetic acid, nor any other acid, has a tendency to favour the development of more acid in the system. I think this should be generally known, for there seems to be a prejudice against the acid wines of France and Germany in this country, as though they were capable of producing the pernicious effects of our own saccharine beers, ciders, and wines.''

In another passage he further remarks,—

"Sugar hides the flavour of acids; so that a sweet wine may really contain much more acid than an acid wine.

"It is well known that gout comes on in Port-wine drinkers. Port wine contains more sugar than any of the wines ordinarily drunk in England. Sugar alone will not produce this disease; but sugar with alcohol, as in Ports and Sherries, will produce it."

Evils attendant on the use of Fortified Wines.

Dr. Maculloch thus writes :—

"Added spirit decomposes the wine. However slow the effects of this decomposition may appear, they are not the less certain. The first and most conspicuous effect is the loss, already mentioned, of that undefinable lively or brisk flavour which all those who possess accuracy of taste can discover in French wines, or in natural wines, and a flatness, which must be sensible, by the principle of contrast, to the dullest palate which shall compare the taste of Claret with that of Port, or

spirits, or ..

this cause we may doubtless attribute the great difference in
the effects produced by an immoderate indulgence in Port and
Sherry, or by a similar use of Claret and other French wines.
Even the immediate effects are sensibly different, as already
said, and the transitory nature of the one, with the permanence
of the other, are too well known to be insisted on. But the
ultimate consequences appear to be of a more serious nature.
It is well known to physicians that diseases of the liver are the
most common and the most formidable of those produced by
the use of spirits. It is equally certain that no such disorders

.

almost every virtue

ose who know no better

found that

taste

He with Sha

and tasting each alternately, they have preferred the Greek. Nay, more; the almost certain consequence of regularly drinking any pure wine is the inability to relish either Port or Sherry; for the flavour of the compounds, arising from added spirit, sugar, &c., becomes thenceforth as unsatisfactory to the palate as they are injurious to the stomach. When tested by such a comparison, Port and Sherry will be found to be deficient in that very principle or vinous flavour which is first sought by true connoisseurs, who are well aware that when certain changes are produced in wines, they are incapable of affording the light and agreeable stimulant which a sound, wholesome beverage provides.

It was one of the most curious anomalies of my early position, that my Greek wines, being without added spirit, should by some have been pronounced poor and deficient in body; whereas their average strength is nearly 26 per cent. of proof spirit, the highest of all natural wines. On the other hand, by many persons accustomed to drink pure French or Greek wines, Greek were thought too strong. By way of commentary upon these opinions, it is worth while to state that the best London porter, and the better kinds of ale, contain but from 10 to 13 per cent. of proof spirit; whereas Greek wines, undiluted, are about twice as strong as such malt beverages; and when they are diluted by one-third water, they are then as strong as French or German wines. So much for the strength of Greek wines.

The following independent Foreign testimony as to the suitability of the soil and climate of GREECE, the character and quality of her wines, and their superiority to French growths, is thus furnished by M. Lenoir, in his work on the

Culture of the Vine, and Wine Making.

"The emancipation of Greece, to which we look forward, will cause a great increase in the culture of the vine, for it there grows luxuriantly, and there is ample room; and when we begin to understand and realize the character and quality of her wines and their suitability for a foreign trade, and that on the vine of all such plants we most rely for support, it will indeed be necessary for us to be on the alert. With such competitors so near to us—competitors endowed with the most favourable climate for vine culture—our wine export, which to-day is in a state of increasing prosperity, will not be long in becoming stationary, and even fall off, if we do not hasten to introduce in the culture of the vine and the mode of wine-making every improvement of which they are susceptible."

Edmund About on the quality of Greek Wines.

The superior quality of the Greek wines, particularly those of the Isle of Santorin, is still further confirmed by Monsieur Edmond About, who states that,—

"The principal agricultural wealth of the country is derived from the vine, of which there are two kinds, viz., the currant and the grape. All kinds of grapes flourish on the soil, and every province produces wine; but the best is from the Isle of Santorin, in which there are more than sixty varieties of the vine. The wines from this island keep a long time, and bear the longest voyages; they please the eye by their brilliant colour, content the palate by their free and fresh flavour, and for the last two years, at meals, I have drunk no other wines."

The above extract from *La Grèce Contemporaine*, printed in 1858, contains an opinion which has been long retained, for it is reiterated in *Le Roi des Montagnes*, published in 1867 :—

" Ce qui contribua surtout à m'acclimater dans la maison de Christodule c'est un petit vin de Santorin, qu'il allait chercher je ne sais où. Je ne suis pas gourmet, et l'éducation de mon palais a été malheureusement un peu négligée : cependant, je crois pouvoir affirmer que ce vin-là serait apprécié à la table d'un roi : il est jaune comme l'or, [the THERA,] transparent comme la topaze, éclatant comme le soleil, joyeux comme le sourire d'un enfant. Je crois le voir encore dans sa carafe au large ventre, au milieu de la toile cirée qui nous servait de nappe. Il éclairait la table, mon cher monsieur, et nous aurions pu souper sans autre lumière."

As regards English opinion, formed on my earliest importations of Greek wines, Dr. Druitt, in his *Report on Cheap Wines,** and their subsequent development and improvement, thus records his experience :—

Dr. Robert Druitt, M.R.C.P., on Greek Vintages and their improvement with age.

" The Greek wines, which I only know through DENMAN, appear to almost perplexing number ; and I believe it would be good policy for the vender to eliminate some of the less important, and fix the public attention on fewer varieties. I first procured these wines in the spring of 1863. I have since studied them with considerable care, and, to say the least, am convinced that they will form no inconsiderable portion of the future wine of this country, so soon as the middle classes, to whom cheapness essential, learn to look out for a decided *wine flavour*,—that is for the taste and smell of the grape, more or less modified by fermentation, instead of the taste of spirits. Nay, more; the specimens I have tasted of some of those wines which have had age in bottle, have led me to believe them so capable of developing flavours of peculiar marked character, that they will be sought out for their own intrinsic excellence, cheapness apart.

"In order to classify them, we may divide them into dry and sweet. Of the dry, there are the *White Mount Hymettus, White Kephisia, St. Elie,* and *Thera;* the *Red Mount Hymettus, Red Kephisia,* and *Santorin.*

"The St. Elie; alcoholic strength$=25_0$. A light-coloured, firm, dry wine; not too acid; clean and appetizing. An older specimen, which had some age in bottle, was a delicious, firm, well-flavoured wine, admirably adapted for dinner. There seems great promise about it.

"So far I wrote in 1865. Subsequent experience enables me to say that the St. Elie went into disfavour with some of my friends from its great acidity and harshness. Blessed is the young wine which has these characters, if only it can be put by to mature. For I find that the St. Elie, if duly allowed to rest, deposits a small quantity of tartar, becomes darker in colour, and acquires a flavour of the true old *winey* character, resembling that of old Madeira. I use the word *winey* to indicate that taste and smell which wine has, and which other liquids have not, and which is developed in the intensest form in this wine.

"Here is a bottle of St. Elie—a white Greek wine, which has been in my cellar awaiting its destiny since 1864. The wine bright, brilliant, dark amber colour. When sipped, one is struck with the feeling of body, the abundance of vinosity: all the organs are gratified, and during the act of deglutition, there is the feeling of power coming into one. The *substratum,* the *fons et origo* of the St. Elie, is a honey-flavoured grape, and this is still detectable; the wine at first was harsh and crude, now it is soft.

"There is another red wine of a peculiar and first-class character, the Noussa, which is said to come from the neighbourhood of Mount Olympus, in Thessaly, which is well worth the attention of my readers who want the maximum of vinosity. It is a rich Burgundy-like wine; soon acquires great flavour, but soon also deposits a crust, and becomes of a tawny colour; sound, dry, not acid, and eminently vinous. The Noussa and St. Elie are the monarchs of Greek wines.

There is a peculiarity of flavour about the Greek wines, resembling that of Amontillado Sherry, and believed to depend on the development of a body called aldehyde, which is alcohol partly oxydized, but not converted into acetic acid.

"The WHITE MOUNT HYMETTUS, 16s., and WHITE KEPHISIA, 20s., as I am informed, differ merely in age. Alcoholic strength about 21°. The White Kephisia is a very cheap wine. It has abundance of wine *taste ;* whereas some that is older has perfectly astonished me by its firm, dry, clean character, and the abundance of peculiar wine flavour of a Tokay sort, which it seems capable of developing.

" The THERA, alcoholic strength equals 25°. This is a wine which, when new from the cask, is of a darkish sherry colour, full bodied, and very capable of taking the place of ' dinner sherry.' Comparing this wine with a cheap fictitious wine of equal price, it is instructive to notice the fulness of wine taste, and absence of spirit taste. The taste is peculiar ; but this wine seems to have great potentiality of developing flavour in bottle. As it is, how superior to cheap Sherry ! White Patras is like Chablis.

" The RED KEPHISIA, 20s., is a wine of great usefulness now, and of great promise when age shall have matured it. The alcoholic strength is about 21°. Full-bodied, dry, markedly astringent, not acid, they are much more satisfying than pure Bordeaux of equal price.

"The SANTORIN is a very useful wine; it has the tawny colour and dryness of light Port, with alcoholic strength about 24° or 25°. I have occasionally given or prescribed this wine to poor patients, and been quite amused at their approbation of it : so like Port ! A dyspeptic overworked dispensary medical officer, to whom I gave some, tells me that it suits him to a nicety, and controls the acidity of the stomach. I have had one or two old samples of Santorin of great merit, as being reproductions of certain characters of old Port. '

" Comparing Greek wines with Bordeaux of equal price, there is more *body* in them, using the word body to imply fulness and rotundity of taste, and what satisfies the stomach, apart from flavour and alcoholic strength also. Persons who

E

might think Bordeaux thin and sour, might be satisfied with
red Kephisia; on the other hand, a person who delights in
light Bordeaux might think the Hymettus coarse, unless he
got some of the older and more mature kind.   Wine flavour, I
need scarcely repeat, is a product of time, and time adds greatly
to the cost of wine; so that in cheap wine we regard not so
much present flavour as firmness and soundness, and capacity
of keeping till flavour shall be generated.

"It follows that the persons to whom we should recommend
Greek wines especially, are those who are hardly weaned from
brandied wine, and who require something full-bodied.   I find
the RED HYMETTUS much relished by a patient in an advanced
stage of phthisis, who says that he really prefers Port, but
that it makes him too hot and thirsty; whereas the Hymettus
quenches his thirst, and gives him 'support' besides.   A second
patient, who has had a narrow escape from puerperal fever,
says it agrees well, and has checked diarrhœa.   The former
patient can afford what he likes; the latter, if she had not the
Greek wine, would have been condemned to South African
Port."

" Of the sweet wines I have not much to say, as they are not
a class of wine that suits me.   The VINSANTO is a very full-
flavoured wine, of very high specific gravity and little alcoholic
strength.   The CYPRUS is marvellously high-flavoured and
sweet; and other wines, as the LACHRYMA CHRISTI, CALLISTE,
partake of these qualities in a lesser degree.

" Sir Edward Barry (1775) says of the Greek wines at that
day, 'that as from their peculiar fine flavour they are not easily
adulterated, they are seldom imported.'   The best of them come
from Santorin, the volcanic isle in the Archipelago, whose wines
long ago were stamped with the approbation of Edmund About,
in *Le Roi des Montagnes.*

The late Dr. Edward Smith, F.R.S., on Greek Wines.

In his recent work *On Foods,* forming one of the
International Series of Scientific Works published by

H. S. King & Co., 65, Cornhill, Dr. Smith records his judgment of new and matured Greek Wines in the following terms :—

"The Greek wines are as yet but little known, far less known than their merits, or the merits of Greece as a wine-growing country deserve; for until a powerful government is established, and there is safety to person and property, the great resources of the country in this direction cannot be developed. No country in reference to soil, elevation, sun, and climate, can excel it, and with capital and intelligence the wines may equal, if not surpass, those of Central Europe.

"These, like all other natural wines, have a character of their own, so that they can only in a very general manner be compared with wines in ordinary use; but all have this characteristic,—viz., that when new they more or less resemble the white and red wines of France or Portugal drunk at dinner; and when ten years old in bottle, have acquired qualities which make them resemble dessert wines, and in flavour not inferior to the most delicious liqueurs. This contrast in quality in the same wine and rapidity of maturation are most remarkable.

"They are produced from grapes which abound in saccharine matter, and in all the elements of grape juice, and being perfectly fermented are probably the strongest natural wines in the market, and, when drunk at dinner, will allow a dilution of one-third to lower them to the strength of the French and Rhine wines. Hence, whatever may be the value of wine, they possess it in the highest degree, and in point of economy cannot be surpassed.

"The following table shows the natural alcoholic strength of the Greek as compared with French wines:—

GREEK.

	Proof spirit		Proof spirit.
Thera	26·00	Cyprus	23·66
St. Elie	26·00	Red Mont Hymet ...	23·40
Santorin	25·92	Red Kephisia ..	23 03
White Patras ...	25·84		
White Kephisia ...	25·63	*Sweet.*	
White Mont Hymet	25·14		
Como	24·54	Lachryma Christi ...	17 13
Red Patras	24·00	Vinsanto	15·61

FRENCH.

	Proof spirit		Proof spirit.
Hermitage	22·03	Sauterne	17·0
Pouilly	21·00	Graves	16·10
Chambertin ⎱ Clos-Vougeot ⎰ ...	20·80	St. Estèphe ⎱ St. Emilion ⎰ ...	16·00
St. George	18·30	Médoc ⎱ Chateau-Lafite ⎰ ..	15·70
Chablis	18·02		

"The new red wines are somewhat astringent, and should be drunk with water, but the white of the same age have a milder flavour, and being without acidity, may be enjoyed equally with or without dilution.

"Of all wines with which I am familiar, none excelled old bottled THERA in the delicacy, fulness, and lusciousness of its aroma and flavour, and being a completely fermented wine made from the fresh grape, it is worthy to be regarded as a perfect wine, and the representative of the Nectar of ancient Greece. It is scarcely possible to make a selection of these wines, which, when young, would be equally appreciated by all persons; yet, perhaps, the WHITE KEPHISIA, ST. ELIE, MONT HYMET, PATRAS, and THERA would be the most generally approved. The St. Elie developes an Amontillado character whilst the Patras more nearly resembles Hock, and the Kephisia Chablis, but with a much greater fulness of body and flavour, Of the red wines, the Noussa, Patras, and Kephisia may be mentioned, all of which resemble the unfortified Rhone or Burgundy wines, and become less astringent when they have deposited a portion of their tartar and tannin by age.

"These full-bodied wines, whether in wood or bottle, de-velope various ethers, and this is particularly observable in the St. Elie. The White Kephisia is a very fine full-bodied dry dinner wine. . . . On the whole, I am of opinion that if these wines should be continued to be prepared by perfect fermentation, without being fortified, and with the body and aroma which they now possess, they must occupy a very high place—perhaps the highest place, among natural unfortified wines; and if the price should be wisely kept down, they must be admitted to universal use, to the yet further exclusion of fortified wines.

" There is yet another class to which reference must be made, viz., the Sweet Wines prepared from the dried grape, of which the Lachryma Christi may be cited as a well-known example. The wine called Vinsanto, a white Santorin, is, however, pre-pared in the same manner, and is a most delicious dessert wine."

HOW TO KNOW PURE WINE.

1.—THAT pure sweet wine is of low alcoholic strength.

2.—That all perfectly fermented wine is dry, and of high alcoholic strength (varying according to circumstances, and then rarely exceeding 26 to 28 degrees of proof spirit,) and is not sweet, as all the sugar from the grape has been converted into alcohol, and should have, when young, an acid, or rather a sub-acid taste (not acetous) from the presence of tartaric acid, which is the natural and healthful acid appertaining only to wine. Its removal by gypsum or plaster of Paris converts the tartrate of potash into sulphate of potash, which is a bitter and purgative salt, with a depressing action on the heart. All sherries are plastered.

3.—That the addition of alcohol to wine, either before or after fermentation, renders it unwholesome, and conduces to gout and similar disorders.

4.—That the greater the amount of natural alcohol produced in wine, the greater is the amount of body in it, as the other constituents of wine must have been produced *pari passu*, and have been existent in the grape to yield the amount of alcohol, whereas added spirit does not give body.

5.—That the greatest amount of natural alcohol in wine is produced in those climates in which the grape attains the greatest perfection, and consequently contains the largest amount of sugar combined with the other ordinary constituents of the fruit.

6.—That as all Port, Sherry, Madeira, Marsala, Catalonian, and Roussillon contain from 36 to 42 per cent. of spirit, they have been either checked in the fermentation by the addition of alcohol to retain the sweetness in the must; or, after the fermentation was completed, the wine must have been sweetened and spirited to bring it - up to the regulation standard. Furthermore, that added spirit causes an undue deposition of the tartrates and neutral salts of the wine, (thereby depriving the wine of that life, freshness, and character which render it so valuable as a remedial agent,) covers defects, and enables all sorts of mixtures to be made up and sold as Port, Sherry, &c.

7.—That the addition of alcohol to wine renders it of less pecuniary value, as spirit costs about 1s. 8d. per proof gallon without the duty, which is much less than the cost of wine; it therefore follows, that if it was not for the Excise and Customs duty, spirit the strength of ordinary Port and Sherry could be sold for 2d. a bottle.

8.—That all natural wines, if any improvement is to be effected by age, must throw down a deposit, and thereby become *sweeter in bottle by the elimination of their tannin, tartrates, &c.* From red wine the deposit contains tannin, which, uniting with the albuminous matter contained in the wine, forms a crust, that year by year becomes less and less, until at length it becomes so thin that it acquires the name of " bee's wing." The deposit also takes the form of crystals, which will both adhere to the cork and fall to the bottom of the bottle like powdered glass. All natural wines that have been any length of time in bottle should therefore be decanted with care.

RESUMÉ.

IF the above data, collected from so many sources, do not carry with them their own effect and propound their own moral lesson, it is not likely that anything that I can add will exert the slightest influence. Yet for the sake of that idea which the great dramatist indulged in, that "a good play is all the better for a good epilogue," I will offer a few remarks by way of conclusion. It seems to me that the recent proposal of Dr. Hassall to "deplaster" Sherry, completes the circle within which the doomed beleagured compounds are being assailed. At first Sherry and Port were combatants of great power and *prestige*, whose supremacy it was almost sinful to doubt, but as their weaknesses became more and more evident, assailants multiplied. Allies plead in vain for a truce, with offers of Specialité and medical certificates, and now we have in Dr. Hassall's grave proposal a plain "Surrender" of all that has been contended for. To change our metaphor, it may be said that the valetudinarian, so long vouched as perfectly healthy, has now, not only been pronounced seriously ill, but an alterative of a very drastic kind has been prescribed for him. Well may the *Medical Times* bemoan "poor unfortunate Sherry," and wonder whether it was worth while going through so much to achieve so little. If œnological science would let Sherry alone altogether there might be hopes of improvement; but between the doctoring which be-plasters, and the doctoring which de-plasters, what worth having can survive? Let us who have watched these endeavours

be ourselves, at all events, wise enough to prefer pure
wine that has no need of either doctor,—fresh, wholesome, palatable, and invigorating,—to the medicated
drinks which should be vended, if at all, by the druggist,
rather than the wine merchant.

Wine drinkers should consult their own health, as
well as their own predilections; and whatever the
country which may supply the wine of their choice—
whether it be France, Germany, Spain, Portugal, Italy,
Hungary, or Greece, whether they drink Bordeaux,
Burgundy, Champagne, Hock, Sherry, Port, St. Elie,
Kephisià, or Lachryma Christi, they should not rest
satisfied so long as one of the chief comforts of life, one
of the principal items of their daily diet is tampered
with and sophisticated, raised in price, and lowered in
value. It is an old maxim that it is of no use disputing
about tastes, but surely there is a necessity for raising
questions of this kind, when common sense, and
medical science and connoisseurship, are all on one side,
and only a very barbarous fancy nurtured on usage upon
the other. The costermonger of the Seven Dials, who
requires his gin made fiery by adulteration before he
can believe it to be " good," possesses in his way what
may be called a " taste ;" and yet few would think it
sufficient to say of him or to him, that " in this damp
climate it is requisite for him to mix his liquor " with
capsicums or with vitriol to fortify him against the
variations of temperature. None would think the
costermonger wise in persuading himself that he liked
such a fearful compound. It is not enough to urge on
his behalf that " he takes what he likes best," and
what to him seems best adapted to his circumstances

and to his palate. We should all join in "disputing his taste;" we should none of us hesitate about drawing the line between gin pure, and gin be-devilled; and yet between wine pure and wine be-spirited we some of us have grave doubts and difficulties.

It is not to be expected that the habit of a lifetime is to be broken in a moment, or that a taste vitiated by long usage is to recover itself forthwith when facts become too patent to be denied. There must probably be a period of transition; when the half-way-house between Port and Claret is found to be some strong Burgundy or coarse vintage of the south of France, or when between Sherry and St. Elie, or white Kephisia the shorter step seems to be a dry and perhaps questionable Manzanilla. Anyhow, it is to be hoped that the next generation will sip Port and Sherry of the liqueur kind, as we do Chartreuse and Maraschino, only occasionally and by thimblefuls. We have lost sight of the Port five-bottle and three-bottle men in the distance of seventy or eighty years. How they survived their potations is a mystery amounting to the miraculous. They are gone at last, and we hope never to see their old fashions revive, nor drunkenness again become a royal, a patrician vice, to be imitated by all the poorer and more ignorant classes. To enjoy and not to exceed is now the general, the almost universal rule; and in order to enjoy thoroughly, the taste must be cultivated, and the mind informed. Without emulating those students of Brillat Savarin, who require a separate wine to savour each remove, there are many who think it worth while to consider the times and circumstances, when white or red, sweet or dry wines, are most desirable. Sometimes

we may choose this, sometimes that, according to our
wants at the moment; but in the whole range of
vintages, from the thinnest *vin du pays* to the most
expensive and most elegant *crus*, there may be ample
pleasure and benefit so long as each is in its way good,
and sound, and natural. The simplest, if pure, is fit
for a king; the richest, if adulterated and compounded,
is not good enough for a peasant.

THE

WINE CULTURE

IN CALIFORNIA.

By HENRY GIBBONS, M. D.

San Francisco:

H. H. BANCROFT & COMPANY.

1867.

The following pages have been written out from the notes of a lecture, which was prepared without any view to publication. They are addressed to the understanding, the heart and the conscience of the people of California. There are three classes of men for whom they are not intended :

1st. Those who estimate the welfare of society by the standard of dollars and cents.

2d. Those who acknowledge no obligation to their fellow men in the way of labor or sacrifice.

3d. Those who live to eat and drink.

WINE CULTURE.

Forty years ago, a few individuals, under the divine inspiration of love for their fellow men, assembled in the city of Boston, to take counsel one of another and to devise some plan of arresting the desolating flood of intemperance. Tracing drunkenness to tippling, the excessive to the moderate use of liquor, and perceiving how difficult it is to cure the habit when once established, they adopted a new mode of warfare, consisting of prevention rather than cure. They directed their labors mainly to the temperate portion of society, in the hope that by inducing them to abstain entirely from ardent spirits, and by training up the rising generation in the same course, intemperance would finally die out. Leading the way by example, and praying for the aid of the Heavenly Father, they formed associations having the pledge of abstinence from distilled liquors as the bond of union. Agents were employed to collect and disseminate information, and the pulpit and the press were enlisted. Temperance Societies, so called, sprang up in all directions, bearing a rich harvest of fruits, and inspiring hopeful and

1*

sanguine hearts with the joyful prospect that they would prove the harbinger of the millenium.

Such was the origin of the great Temperance Reformation, which in a few years spread its vivifying influence from the new to the old world, everywhere dispensing rich blessings, and everywhere hailed as a precious gift of God to man.

But the good work soon came to a stand. Drunkards who had been redeemed began to slide back into the fiery tide from which they had been rescued ; and another crop of drinkers and drunkards began to appear in the rising generation. Wise heads traced the failure to the continued use of fermented drinks, which had not been considered dangerous. The notion had prevailed that ardent spirit was the great enemy— that distillation created the poison—that the worm of the still was the serpent of intemperance. To substitute wine and beer for stronger drinks, had been the key-note of reform. Train up the youth to indulge moderately in these beverages, and you will banish intemperance!

And now was sounded the tocsin for a war which was destined to revolutionize public sentiment. That the habit of intemperance once formed could not be cured by anything short of abstinence from all intoxicating beverages, was a truth, plain, palpable, undeniable. Equally beyond all controversy was the proposition that the moderate use of any kind of intoxicating drink tended to the immoderate use. The radical cham-

ions of temperance determined to enforce these
·uths by example and precept. They took coun-
·l of an ancient hero, who had proclaimed the duty
f abstaining from all indulgences through which
is brother was made weak and caused to stum-
le. They saw their brethren stumbling and
·aggering through wine ; and they nobly pledged
·emselves before man and before God to drink
o wine while the world should stand.

The battle was joined. Legions of the old
·mperance men, faint hearted, thirsty or mam-
·on-struck, left the ranks and deserted to the
·nemy. The advocates of total abstinence were
·enounced as ultraists and fanatics : even as infi-
·els, who trampled under foot the Word of God.
·Iinisters of the Gospel transformed themselves
·nto priests of Bacchus, and invested the luring
owl with the hallowing associations of religion.
·he Bible, with reckless resolve, was thrust for-
·ard to cover the sparkling cup ; and the broad
·ay to indulgence and drunkenness, was kept
·pen in the name of religion. The weapons of
·ne party were furnished by the heart and the
·onscience; of the other, by the stomach and the
·urse—by the world, the flesh and the devil.

In that great struggle was witnessed what the
·istory of human progress has more than once
·evealed—that faith and devotedness and unwaver-
·ng perseverance were on the side of the radical
·eformers : and that the victory too was theirs.
·n a few short years they had exclusive and un-

disputed possession of the field; and since that
triumph, temperance has had but one signification
all over the civilized world, namely: THE MODER-
ATE USE OF THINGS USEFUL, AND TOTAL ABSTINENCE
FROM THINGS HURTFUL.

And now, here in California, history repeats itself, and the old battle is to be fought over again.
Actuated by the prospect of gain, a motive legitimate in itself, large numbers of our most influential citizens are engaged in a systematic attempt
to make California a wine-producing and wine-
drinking country. One single motive is at the
bottom of the movement. No one will pretend
that the promotion of temperance is the object, or
a prominent object, of our wine-culturists. It is
politic to urge the point, for the purpose of removing objections and facilitating the main de-
sign. They petition Congress to take off the duty
on domestic wines, that they shall be made cheap,
and brought within the reach of the people as a
common drink, and thus promote temperance by
crowding out brandy and whisky, which are con-
ducive to drunkenness. Having accomplished
this much on behalf of temperance, their next
step is to get the duty removed from domestic
brandy; and thus brandy, also, is to be cheapened
and brought within the reach of every one. The
first stand point is temperance; and when that has
served its purpose, they trip with light fantastic
toe to the opposite stand point of—*brandy*.

Let me say here, for myself and those who are

with me on this question, that in our regard for
the interests and welfare of California, we are not
to be placed in a position subordinate to any
other class of our fellow citizens. Here is our
chosen home, and here is the home of our children. Here we and they are to earn our daily
bread, and to fight the battle of life. To make
this a better and a safer home for our children,
and for the coming generations, we have taken the
vow of the Nazarite, we have made the pledge of
the sons of Rechab. And it is for this we are assembled here to-night—not in the spirit of strife
and hatred, but in all love and kindness, to plead
with wine-growers on behalf of their own offspring, and to urge the superior importance of intellectual and moral interests in comparison with
material welfare. If the wine-growers are right,
their success will add wealth to the State, at no
expense of life and morals. But if *we* are right,
their success will bring to future generations a
harvest of fire-brands, arrows and death!

It is important to bear in mind that there is but
one source of alcohol—the intoxicating principle
contained in liquors—that the one source is the
process of fermentation—that distillation only separates the alcohol from its combinations in fermented liquors—that distillation never created an
atom of alcohol—that not a particle of alcohol exists in nature, or was ever produced by the hand
of the Creator—that it is the product of the decomposition of sugar—that when the juices of

fruits containing sugar in abundance, and which
are perfectly harmless and unintoxicating, when
first expressed, are suffered to stand for a time in
a certain temperature, the finger of death touches
the liquid and destroys the sugar, converting it
into alcohol, the intoxicating or *poisoning* principle—for such is the literal meaning of the word.

When grains are used instead of fruits, the
starch which they contain is first changed into
sugar by heat and moisture, and the sugar is
changed into alcohol by fermentation. But in
every instance the alcohol must first be formed
before one drop can be procured by distillation.

With these facts in view, let me ask this intelligent audience to consider for a moment the difference between fermented and distilled liquors.
From a standard medical authority I take the
following enumeration of the components of wine,
as exhibited by chemical analysis: œnanthic ether
and œnanthic acid, on which depends the "bo-
quet"; carbonic, acetic, tartaric, malic, and tannic
acids; bi-tartrate of potash; tartrates of lime, of
alumine, of iron; chlorides of soda, of potash, of
lime, of magnesia; sulphates of potash and of lime;
coloring matter, and sugar.

Remember, this is the composition of *pure* wine.
When wine thus constituted is subjected to distillation, the alcohol is driven off unchanged, in
combination with a small portion of "fusel oil"
and of coloring matter, whilst all the acids and
salts are left behind in the still. Distillation

therefore simply separates the alcohol from what may be deemed the impurities of the wine, and isolates and concentrates it. Thus brandy consists of alcohol combined, half and half, with water, whilst wine consists of the same element combined with from five to ten or twelve parts of water, and holding in solution the various articles above enumerated.

So with all fermented drinks—they all contain alcohol. In order to drink a gill of alcohol, one must swallow from three to four pints of cider, beer, ale, porter, or the weakest wines, one pint of the strongest wine, and half a pint of fourth-proof brandy, whisky, or spirits.

Now what becomes of alcohol when taken into the stomach? Articles of food are slowly dissolved by the gastric juices, and then passed out of the stomach into the small intestines, there to be still further altered and digested, so as to form the chyle, which is taken up by the proper vessels, and converted into blood. But alcohol undergoes no such change. Scarcely has it entered the stomach before it begins to pass into the veins, and in a few minutes it is coursing through all the channels of circulation, bathing with its fiery flood the liver, lungs, heart, brain, and every organ and fibre of the body. By its direct contact with the brain, it produces excitement and intoxication. The lungs detect the poison in the blood, and they pump it out in the form of vapor in the breath. If the supply be frequently renewed, it lodges permanently in the important organs, especially in the liver and brain. It has been found in the brain sufficiently concentrated to blaze on the application of a lighted taper. Everywhere it tends to kindle disease. And it has the peculiar power, beyond any other known agent, to deprive the individual who drinks it of the ability to perceive its effects either on his mind or his body.

The medicinal use of alcohol, so far from invalidating this argument, serves to confirm it. Whatever is active as a medicine must be hurtful in health. Men do not take strychnia, or arsenic, or opium when in health because these articles may do good in disease. Ipecacuanha or castor oil may restore health to the sick; but he who, for this reason, should swallow them when not unwell, would be accounted a fool. Besides, the use of alcohol in health cuts off the drinker from whatever benefit he might derive from it as a medicine. To temperance men it may be a medicine, but to no others.

Now there is this difference between distilled and fermented liquors : the former are more rapidly absorbed, and their effect is more speedy and more evanescent; whilst wines and other fermented liquors, entering the blood more slowly, their intoxicating influence is proportionately less, and the disturbing consequences of greater duration. Besides, and this is a matter of importance, whilst brandy and whisky act almost entirely on the nervous system as stimulants, wine has an additional action on the stomach as an irritant by virtue of the acids and extraneous substances with which the alcohol is combined. That is, whilst the brain chiefly is poisoned by the former, both brain and stomach suffer from the fermented liquor. For this reason many persons can drink spirits in moderate quantities with comparative impunity, whilst they cannot tolerate wine. And for the same reason, the debauchee suffers more the next day from wine than from brandy. For this reason the temper suffers most from wine, which makes some men irritable and devilish, whilst whisky makes them foolish and stupid.

The presence of so many extraneous substances in fermented liquors tends to prevent their keeping, and to induce a second fermentation by which the alcohol is transformed into vinegar. On the other hand, distilled liquors are incapable of such change, and will keep indefinitely. For this reason brandy is added to wine, in order to prevent its souring.

But the expression of the juice and the fermentation of it, constitute but a portion of the art of wine-making. It must be prevented from souring; inferior qualities are made saleable by adding better; brandy or alcohol is added to thin wines to give them strength; acid wines are treated with sugar, honey and raisins, or the acid is neutralized with potash, soda, or lime; the paler kinds are improved in color by burnt sugar. Books are written on this subject, laying down the rules. Not always such innocent means are resorted to.

Sugar of lead, a well-known poison, has been extensively employed to impart the proper astringent taste. An English manual directs two and a quarter pounds of small shot to be thrown into each cask, to prevent souring. Carbonate of lead, an active poison, is thus formed in the liquor. In extreme cases, when other methods have failed, it is advised to throw into each bottle a pinch of oxalic acid, a virulent poison.

Then comes the manufacture of factitious wines, which goes hand in hand with the wine-trade, all over the world. To such perfection has this branch of the wine manufacture been developed, that good judges have frequently taken for the best genuine wine, an article that did not contain a drop of grape juice. The port and champagne of commerce are nearly all factitious. Brandy or whisky, or alcohol itself, is used as the basis, in some cases cider and gooseberry juice, and an immense variety of ingredients are employed, such as alum, logwood, oak bark, willow bark, rhatany root, litmus, red saunders, beet juice, &c. A single recipe for an imitation wine has sold for four hundred dollars in the city of New York, and fifty dollars has frequently been the price of an empty cask stamped with the European brand.

Perhaps there is no traffic in which men ever extensively engaged, so universally corrupting to those who pursue it, as the trade in intoxicating liquors. In all times and in all countries it has tended to demoralize the dealer and to defile his character. I speak not of the effect of drink on the vender, at whose feet the dark and ghastly pit which he digs for his customer yawns evermore. My allusion is to the frauds which have always been a part of the history of intoxicating liquors —frauds identified with their manufacture and sale. There seems to be no traffic on earth which presents so many and such strong temptations to dishonesty. Strumpf, a German authority, quoted by our standard authors on materia medica, declares:

"The wine trade is, with few exceptions, in the hands of avaricious usurers, knavish dealers, and greedy landlords, who have more regard for the gratification of their own avarice than for the health and lives of men. French, Spanish, and all other foreign wines, seem without exception to be artificial, rather than natural productions. Still worse, even German wines seldom form an exception to this statement, and it were easier to obtain pure musk or genuine cinchona bark than faultless wine. Under such circumstances it is a precarious matter to attempt the cure of our patients with a liquid which is hardly ever what it ought to be."

Pliny made the same complaint 1800 years ago: "If all were agreed which is the best wine, who could get it? Our very princes do not drink pure wine. To such a pitch has villainy arrived that one can buy nothing more than the name of a vintage. From the very wine-vat all of it is adulterated. And so, marvelous to tell, we may say of wine, the poorer the purer."

But our California wines are to be pure—unadulterated. For the first time in the history of wine making, the intoxicating cup is to be flavored with honesty. I will not deny that Californians are the most conscientious and scrupulous people in the world in matters of traffic; that they are capable of resisting temptations which sweep the rest of mankind into knavery. But I read in the proceedings of their wine-growers' convention that pure wines will not suit the popular taste—that they require a little legerdemain or something else to make them sell; and that eastern people are complaining already that California wines contain too much brandy. And this in the green tree!

We have had a swill milk panic in San Francisco. It was discovered that certain milkmen were feeding their cows on the refuse of distillation—the mash remaining in the still after the alcohol was driven off by heat. To prove this mash or swill to be wholesome food for cows, the milkmen endeavored to show that all the alcohol, except about $\frac{1}{4}$ of one per cent. had been driven off; and they argued that so small a quantity could not hurt the cows or injure their milk. But there was too much humanity among our citizens to tolerate such an idea. It was known that for some reason the swill greatly increased the quantity of milk—that the milk was rendered thinner and less nutritive. It was charged that the children of the poor, to whom this poor milk was fed, were often sickened by it, and some of them died in consequence. In vain the dairymen subjected their milk to chemical analysis to establish its purity and richness—in vain they exhibited the cows as models of health—in vain they proved that there was no appreciable amount of alcohol left in the swill to poison the cows—in vain they argued that there was nothing in the swill but the constituents of the grain that had not been changed into alcohol by fermentation. The guardians of the public weal were inexorable, and they cried out in a voice of thunder—PROHIBITION! And now, cows shall no more be fed on still slops. The public good requires this restriction.*

But what becomes of the alcohol? In the mash before distillation, you had a large portion of alcohol, mingled with the debris of the grain. You have separated the alcohol—the really active and poisonous element—except a trifling per centage, too much, however, to render the refuse fit to be fed to cows. The concentrated poison is now to be fed to men, women and children. The fiery stream is to be poured down human throats, and injury, disease and death will follow. Every morning, there shall stand at the bar of justice a score of victims, ragged and bruised, to attest the

---

* The action here referred to was subsequently annulled by a decision of the Court, declaring the ordinance illegal. But this result does not affect the force of the illustration.

effects of something worse than bad milk. Penury, gaunt and haggard—homes once happy now deluged with lava from hell—debauchery and riot and murder—all these shall curse the people. A few humane individuals, gathered from various quarters of the State into a temperance convention, contemplate this fiery deluge and suggest that it claims the concern of the guardians of the public weal, as much as the diminutive streamlet of swill milk. Whereupon the press, the liberal, enlightened and moral press of the city, which has so boldly applied one edge of its impartial sword to the throats of two or three poisoners of cows, turns the other edge against the temperance convention, to defend and protect two or three thousand poisoners of men! *

But what has this to do with the wine question? A great deal! Before the wine culture loomed up as a great public interest, the press was a unit on our side. Now, the wine men publish the boast that it is a unit for them. No man can serve two masters.

The inquiry into the effects of still slops on cows is a good thing. The animal yields unwholesome milk, and the milk sickens and often kills the children who use it. And what effect have still slops on women—on the milk of the human mother?

---

* This has more special reference to the *Evening Bulletin*, which published an editorial grossly misrepresenting the action and designs of the State Temperance Convention, and then refused to insert an authorized statement denying the charges and setting forth truly the principles and purposes of the Convention.

And when I say still slops, let it be distinctly understood that I place all alcoholic or intoxicating beverages on the same basis, and so far from making a distinction in favor of wine or fermented drinks, were I to discriminate at all, it would be in favor of the purer distilled liquors. The intoxicating principle enters the blood through the stomach, courses through all the channels of circulation, and pollutes with its noxious presence the fountain from which the infant draws its life. I have seen a child drunk from its mother's milk. I have seen another child, within the last month, die from the effect of poison milk—not the milk of cows fed on swill—but of a nurse who drank swill. Let mothers consider these things. Let them reflect that when they take alcohol into their stomach, no matter in what form, whether undisguised as brandy or whisky, or in the more covert, insinuating and deceptive shape of beer or wine, they are feeding it to the infant at the breast. And let every mother in the land who pursues this practice, hear the cry of *swill milk* ringing perpetually in her ears by day, and disturbing her slumbers by night.

And this brings into view the influence of alcoholic beverages on individual and national character, and the consequences on progeny, transmitted from one generation to another. It will scarcely be doubted by any one, that, of all substances capable of entering the human stomach, no one has the same remarkable tendency to animalize and brutalize the man—to develop the lower and animal character at the expense of the higher moral faculties. By an inexorable law of nature, such effects are transmitted to the offspring. Dr. Howe ascertained that of 300 idiots in the State of Massachusetts whose history could be traced, 145 were the offspring of parents one or both of whom were intemperate. In a single family where father and mother were both drunkards there were seven idiotic children!

But these are extreme results, growing out of manifest drunkenness. Moderate drinking has its proportionate influence. If the same laws, and the same care were applied to the rearing of men as our farmers apply to the improvement of the breed of horses, or cows, or even hogs, alcoholic drinks in every form would be banished forever.

Nearly a century since, this subject attracted the attention of Smollett. "It must be owned," wrote this eminent historian, "that all the peasants of France who have wine for their ordinary drink, are of a diminutive size in comparison with those who use milk, beer, or even water; and it is a constant observation that when there is a scarcity of wine the common people are always more healthy than in those seasons when it abounds. The longer I live the more I am convinced that wine and all fermented liquors are pernicious to the human constitution; and that for the preservation of health and the exhiliration of the spirits there is no beverage comparable to simple water."

So wrote Smollett in 1776. In 1866, ninety years later, a French physician, M. Joly, who has devoted much time to close investigation of the subject, declares in so many words: "An increasing tendency towards mental disease has been generated by the increasing consumption of alcoholic drinks, and in proportion as liquor drinking increases, so do paupers, vagabonds, beggars, suicides, idiots, dwarfs, and madmen increase. A hundred years ago France only consumed 200,000 barrels of liquor yearly. She now consumes four millions of barrels." Dr. Joly also maintains that of late years mental maladies have made a fearful progress. An official report prepared in 1865 fully bears out his theory that dram drinking is one of the principal causes of the evil.

Long before this, the wise and good Fenelon had addressed the voice of warning to Louis XV. You ought never — thus he counselled—"You ought never to allow wine to become too common in your kingdom. It causes diseases, quarrels, sedition, idleness, aversion to labor, and family disorders."

Pliny the younger, one of the purest men of his age, has left on record a graphic picture of a wine-drinking people, which may be pondered with instruction by all such advocates of the wine culture in California as can be influenced by moral considerations. He complains that his countrymen bestow great pains and expense on a liquor that deprives man of reason, renders him

furious, and is the cause of an infinite variety of
crimes; that they had invented ninety-five different
kinds of wine, and perhaps double that number;
that they even resorted to the steam bath to increase their capacity for wine; that they vomited
and drank anew; that gambling, lewdness and prostitution were associated with wine-drinking; that
many perish in consequence of words uttered in a
state of inebriety. He tells us that the art of
wine-drinking had its laws; that Marcellius Torquatus, the pretor, observed these laws exactly
in never stuttering nor vomiting at table, in remaining till the morning at his potations, in never
taking breath nor spitting whilst he drank, and in
never leaving in his glass any heel-taps which
could produce sounds when thrown on the floor.
In the reign of Tiberius Claudius it became the
custom to drink wine in the morning on an empty
stomach—a custom he says, introduced by certain
physicians who wished to make themselves popular—some of whose descendants, let me remark in
parenthesis, have emigrated to California. Pliny
also furnishes some notable instances of wine-
drinking. Mark Antony was a drunkard, and
even wrote a treatise in defense of his habit.
Torquatus, already mentioned, was surnamed Tri-
congius or the three gallon knight, because he had
once drank three gallons of wine at a single sitting
in the presence and to the astonishment of the
Emperor Tiberius. Lucius Piso obtained from
the same prince the prefectship of Rome by sit-

ting at the table with him for two days and two nights in succession. And Marcus Tullius Cicero, the hero of our classics, swallowed two gallons of wine at a draught, and in a fit of drunkenness threw a glass at the head of Marcus Agrippa!

Such is Pliny's view of the inner life of the great Roman Empire, in its Augustan age. Cicero himself tells of dining, by invitation, with the Emperor, when the latter paid him the regal compliment of taking an emetic before dinner, to increase his capacity for wine.

But, in our day, wine is no more wine! It has ceased to be a poison! Instead of producing drunkenness, it inspires sobriety! If we can only succeed in pouring it down the throats of all the men, women and children in California, intemperance will vanish—especially if the tax on domestic brandy be abolished, so that our wines may be distilled and converted into brandy! Are not all modern wine-drinking nations remarkably temperate? If a congressman should doubt it, Ross Browne will send him a basket of the "pure juice."* If an editor, a dozen bottles of Angelica will answer. So shall their eyes be opened: and in their reconstructed vision Bacchus himself will be transfigured into a Saint!

Some Americans have a fashion of abusing their

---

* In his report to the Wine Growers' Convention of California, Ross Browne boasts of having convinced the committee of Congress of the propriety of reducing the tax on native wines by sending them a liberal supply of the article.

countrymen as a nation of drunkards, and laud-
ing the nations of Europe as models of sobriety.
They see nothing but drunkenness at home and
nothing but temperance abroad. Among the
French, the Italians, the Swiss, the Germans,
the Austrians, all is temperance and self-control;
among their own kinsmen, all is wassail and revelry. The people of wine-making countries have
been so long held up as patterns in this respect,
and the American people so long belied as drunken
vagabonds in comparison, that thousands of persons honestly believe it. The testimony of travelers is adduced in support of the error. Let us
give a few moments to the consideration of this
testimony.

The moral condition of a community is not
easily penetrated by a casual traveler, and his
impressions in this respect are apt to be founded
on accidental glimpses. A stranger might spend
a week of diligent exploration in San Francisco
without crossing the path of an inebriated man.
He might go with his friend and visit the Mercan-
tile Library, the Hall of Pioneers, the Orphan
Asylums, the Theatres, might attend an immense
political meeting at Platt's Hall, might tread the
thronged highway of Montgomery street again and
again, and come to the conclusion that we are the
most sober people in the world, and so write us
down. Another might stroll along the Water
Front or Pacific street about midnight, or drop
into the Police Court on Monday morning, and

make up his mind that we were the most drunken people on the continent. Besides, men are apt to see what they look for and to confirm preconceived notions by observation. Then again there is no sharp line between sobriety and drunkenness. The Emperor Maximin could carry sixteen bottles of wine at a draught without staggering; and in the estimation of many excellent judges no imbibable quantity of lager will intoxicate. Further than this, a traveler might possibly swallow enough wine to affect his jugdment. I have known a whole party of a score of "gentlemen" to get drunk together, and yet not one of them but would swear honestly that they were all sober.

The value of negative testimony in regard to popular drunkenness may be curiously exemplified after a "Fourth of July," or any general holiday, by asking of different individuals the result of their observations on the subject. Some will felicitate themselves on the remarkable prevalence of sobriety, whilst others will declare that they never saw more drunkenness. Last summer, a friend described to me in glowing colors the decorum and strict sobriety observed by the Germans in their Sunday pic-nics at Alameda. It so happened that on the previous Sunday I had returned from Alameda on the same boat with such a party, and was forced by the crowded condition of the boat to take my place on the lower deck near the bar-room. And certainly I never in my life witnessed a more flagrant and disgusting bacchanalian exhibition than on that occasion. At the very same time, persons on the cabin deck, in the same vessel, mixing with another portion of the same crowd, saw nothing but strict order and decorum.

In a late number of a San Francisco paper is a letter from a correspondent in Vienna, which shows very forcibly how sobriety and beer-drinking may go together. The writer describes the beer of Vienna as a most delicious article, which "leaves upon the stomach none of that nasty, medicinal taste, which everybody who has drank the lager beer of the United States over night will remember to have experienced the next morning. Everybody, man woman and child, drink it. It is in fact the common beverage of the people. No one but a madman would think of drinking water, the only use of which in Vienna, beside a little that is 'wasted' for washing and culinary purposes, is to make beer of. Scattered all over Vienna are immense beer houses, where in the evening everybody goes, whole families being seated together at the table. It is a queer sight, one of these Vienna beer-houses, about ten o'clock in the evening. From 500 to 1000 people, enveloped in a thick atmosphere of tobacco smoke, are seated at the table, eating, and drinking beer."—"You enter and take a seat, and in a moment an immense beer flagon, a 'krugel,' holding nearly a quart of the delicious foaming nectar, is placed before you. The waiter does not ask you if you want it. He takes it for granted that you are a 'man and a brother,' and of course you want beer. Neither does he wait, when your glass is exhausted, for you to call for more; the moment you have set it down after draining its contents, he seizes it and soon returns with it renewed."—"In this way in the course of an evening, three, five, seven, ten or a dozen krugels, or from half a gallon to two gallons, are got through with. It is astonishing how much beer a man will hold. My curiosity is not great; in fact, in the matter of stomachic capacity I would be a "light weight." But by dint of practice which however did not require much effort or self-sacrifice, I can manage to stow away my four krugels (nearly a gallon) of an evening and feel none the worse for it the next morning."

And this is temperance! This is a specimen of European life, and European training, which tourists are so fond of extolling, and offering as an example to Americans! In the same paper which contains the letter above quoted, I find the following scrap. The letter is dated June 2d:

"Two SUICIDES A DAY IN VIENNA.—Last April there were 65 attempts at suicide in Vienna. 47 were by men, 15 by women, 13 by children from 9 to 14 years of age. 22 persons hanged themselves, 15 drowned themselves, 11 took poison, 5 cut their throats, 2 shot themselves, and 7 died of self-inflicted stabs."

In 1839 Edward C. Delavan, of Albany, visited Europe for the express purpose of ascertaining the effects of wine-drinking on national character.

2

On his return, whilst in England, he addressed a
large meeting in Exeter Hall, and made use of
this language: "I have recently taken an extensive tour in the wine countries, for the purpose of
informing myself on this subject, and I had the
favor of a long interview with the king of the
French—the king of that country which produces
more wine than any other on the face of the earth.
And what was the result of that interview? He
said that wine was always injurious and never
beneficial, and that the drunkenness of France was
drunkenness on wine. I put the question to him
a second time and obtained a similar answer.
And his son, the Duke of Orleans, told me he entirely agreed with his father; and he added these
additional facts; that where wine is produced
there is more pauperism, more crime and more
trouble than in any other part of the kingdom;
and that such is the avidity with which the
land is cultivated for this purpose that there is
actually becoming a scarcity of cattle. In trav-
eling through the wine districts," adds Mr.
Delavan, "this opinion was confirmed; I found
it extremely difficult to get on, on account of
beggars."

An English gentleman residing in Paris twenty
years ago, wrote: "The canaille of Paris in 1840
were estimated at 63,000 persons, who were addicted to every variety of vice and crime. But
there was one connecting link which bound them
all together—they were all drunkards, and one-

half of them were brutally drunk. At the same time it was ascertained that there were 20,000 women in Paris who were notoriously given to drink, and 10,000 of these were if possible more abandoned than the men."

This may appear exaggerated. But about the same time a French writer, M. Frazier, makes this statement: "In reviewing the habits of the workmen of Paris, we have pointed out a vice which has the effect, not only of degrading their character as men, but of consuming in a wasteful manner a large part of their wages, and thus depriving their wives and children of the necessaries of life. This vice is drunkenness. It is spread through all classes of the workmen. Those who work in factories are especially addicted to this vice."

Robert Walsh, a name well known in literary and political circles, who resided some years in Paris, wrote in 1838: "The laboring classes in Paris can be called more sober than the British or Americans, only because the number of them is relatively small who use spiritous liquors and who get drunk in our sense. But they drink an enormous quantity of small wine. Tippling shops are everywhere and are abundantly frequented. The wine makes them tipsy or heavy."

J. Fennimore Cooper, the American novelist, makes this important statement: "I came to Paris under the impression that there was more drunkenness among us than in any other country, Eng-

land excepted.   A residence of six months in
Paris changed my views entirely."

Rev. Dr. Hewitt, who visited Europe as agent
of the American Temperance Society, makes this
report: "The common people of France are
burnt up with wine, and look exactly like the
cider-brandy drinkers of Connecticut and the
New England rum drinkers of Massachusetts.   If
they do not drink to absolute stupefaction or intoxication, it is because sensuality with Frenchmen
is a science and a system.   They drink to just that
point at which their moral sense and judgment
are laid asleep, but all their other faculties remain
awake."

Similar is the testimony of Rev. E. N. Kirk,
founded on sixteen months of observation: "The
conviction produced on my mind by all I saw, is,
that no nation is more injured than France by the
use of alcoholic drinks, in regard to the health,
character, intellectual and moral progress of the
people, and also as respects their political and
pecuniary interests.   I never saw any other people
who think that coffee strong to blackness and un-
relieved by milk, is too weak, and therefore fortify
it with brandy.   I never saw the poorer classes of
any other city than Paris, so regularly and in such
immense numbers (some of them remaining from
Sunday noon till Tuesday morning, I am told) at
the places of drinking.   I have inquired among
the most intelligent of the common people, concerning the effects and the extent of the effects of

their beverages; and the answer often seemed like hearing one of our temperance speeches. I fully believe that some things called characteristics of France would disappear with the disuse of alcohol. There is a great deal of ingenious drinking in Paris. Many a constitution is impaired, many a bad passion inflamed, while the discreet wine-bibber has learned just when to stop and to preserve self-command. I have never seen more drunken men in the streets of any city than in Paris; while my belief is that the police is the most vigilant in the world in the prevention of such exhibitions."

And again, the same writer says: "I never saw such systematic drunkenness as I saw in France during a residence of sixteen months. The French go about it as a business. I never saw so many women drunk."

The witnesses whom I have examined were not transient travelers, rushing through on the gallop. They were temporary residents, and besides, they studied the habits of the people. Others tell a different story. Mr. Eddy visited Paris in 1851, and "did not see a drunken man, or witness by night or day, one scene of disorder." He attributes this, however, to the vigilance of a remarkably efficient police. Dr. Walter Channing, of Boston, made the tour of Europe hastily ten years ago, and "saw no drunkenness on the continent"—not even in Russia. He heard of drunkenness in Paris, outside the barriers, where liquors are free

from duty.   But he says people must get sober
before returning, as he saw none in the city proper.
He found the Bavarian "true to his stomach"—
with "his lager ready at all hours and everywhere."   And yet he "did not see an instance
of drunkenness in Germany"!   He saw but one
drunken person in Russia.   And yet it is notorious that intemperance abounds in Russia, the
government deriving a vast revenue from the immense amount of whisky consumed.   Indeed, Dr.
Channing informs us in another place that the
Russian laborer will work the whole week, and go
to the drinking places on Saturday evening, and,
it may be, drink all night.   He describes the din-
ners at Copenhagen, where "to eat generously
and.to drink much wine is the order of the feast."
* * *   "The effect of wine declares itself," he
adds: "the liberty of speech is certainly en-
larged."   But everybody in Denmark was strictly
sober!

The habit of constant, daily swilling, though
not pushed to the destruction of self-control, is
really the worst form of intemperance.   The German who is always stuffed with beer, the English-
man with ale, the Frenchman with wine, the
American with whisky, thus keeping up a per-
petual strain on the powers of life, though reputed
sober, do greater violence to their physical and
moral nature than the acknowledged drunkard who
breaks out in paroxysms and then refrains and re-
cuperates.   Your steady drinker, who cannot swal-

low a meal without potations of alcohol, and who retains, with the appearance of robust health, a reputation for sobriety, is always ready to topple into the grave, either by a shock sudden as the thunderbolt, or by the invasion of disease which a sound constitution might escape, or at the most endure without peril. No class of men on earth appear more healthy and robust than the beer-drinking porters and employes of the London breweries, but they have long been known as the dread of surgeons, for the reason that they are incapable of resisting injuries like other men, and often yield their lives to the effect of a slight bruise or scratch.

And what shall be said of the curses entailed on his offspring by the parent whose blood is forever poisoned by alcohol! Or who can estimate the human monstrosities engendered by the beer-vats of Bavaria, the wine-presses of France, the gin-palaces of England, and the saloons of California!

But I must hasten to present a few additional items of testimony, selected from the mass which has accumulated on my hands. Judge Acton, the superior of all the criminal courts of Rome, wrote to Mr. Delavan, in 1839—"I beg leave to state my opinion upon the proportion of crimes which in this country may be traced to the immoderate use of wine, or to the too great frequenting of public houses. I think I may fairly reckon one-third under that head." "In my last conversation with Judge Acton," says Mr. Delavan, "he stated that only the night before, a man had drank to intoxication at one of the wine shops, gone home and butchered his wife and mother."

That was in 1839. During the last year a volume has appeared from the pen of the Rev. C. M. Butler, D. D., who resided lately some months in Italy, and who made the condition of the people an object of special observation. I quote from this volume: "We have heard Americans assert that there is no drunkenness in any country where wine takes the place of stronger liquors. Now we have sifted this matter thoroughly, both in Italy and Switzerland, and are bound to deny the truth of the statement. Why is it that so little drunkenness is seen by strangers? Because Italian laborers rarely begin their potations until their day's work is done. They carouse from about nightfall to midnight, when, their money spent, or credit exhausted, they reel home, and the cries and groans of wives and children, soon tell of the fury and brutality which mark the drunkard the world over. Thinking it probable that brandy did most of the mischief, I inquired as to this point. In every case my question caused surprise, and the answer was always the same: 'No, no, it is wine—always wine.'"

I must not omit the testimony of a gentleman who is well known to Californians as a former resident of San Francisco, the Rev. Mr. Lacy. Here are his words:

"The testimony of travelers in Europe, as far

as I ever heard, was to the effect that intoxication was very little known in wine-producing districts, and that if wines were only cheap and unadulterated in America, the vice of intemperance would be greatly abated, if not entirely removed. I was so well convinced by such unanimous testimony, that I regarded the introduction of the wine culture in California, and its general increase, as a harbinger of public good, and as a kind of assurance of general morality. I have just spent six months in a country place of Switzerland, where the people do nothing but work in their vineyards; where wine is cheap and pure, and far more the beverage of the laboring classes than water; where none think of making a dinner without a bottle of wine; where all the scenery about is of the most elevating and ennobling character. Here more intoxication was obvious than in any other place it was ever my lot to live in. * * * On holidays and festal occasions, you might suppose all the male population drunk, so great are the numbers in this deranged or beastly condition."

Looking still further east, from the very borders of the Holy Land we, have the testimony of Rev. Mr. Perkins and Dr. Grant, American missionaries to Persia, in 1837. "A great quantity of wine," say they, "has been made the past season, and the consequence is there is at this time the most appalling prevalence of intemperance." "Intemperance is a mighty evil in this country among the

Nestorians. It is said to be less prevalent among the people back in the mountains where the vine is little cultivated."

The Rev. Mr. Schauffler, missionary at Constantinople, writes : "The German colony at Odessa, on the Black Sea, planted vineyards, and in the second generation became excessively intemperate from their products."

I shall refer only to one other authority, that of Horace Greeley. In Savoy, France, where wine is the only article of export, he saw women at work bare-headed, in the fields, hoeing corn, driving cows, hogs, etc., and a girl of fourteen years driving an ox cart. Such rough, hard work, he says, is fatal to every trace of beauty. He saw not a woman in Savoy even moderately good looking. In Switzerland women and girls were breaking stones at the road-side. Lombardy, which ought to be the richest country in the world, was filled with beggars. The dwellings were shabby and the barns scarce. Cattle were treading out the grain under the open sky. And this, he exclaims, is the garden of sunny, delicious Italy. "Italy, beauteous, bountiful land, is everywhere haggard with want and wretchedness." "Brown, bare-headed, wretched looking women, were washing clothes in the hot sun at the sea side. Along the Mediterranean the road runs through a naturally fertile and beautiful country, once densely peopled and covered with elegant structures, the homes of intelligence, re-

finement and luxury. Now not a garden, scarcely a tree, and not above ten barns and thirty houses or habitations for a distance of twenty-five miles. Such utter desolation and waste in a region so eligibly situated, can with difficulty be realized without seeing it. "

But suppose it be true that there are wine countries in Europe where the people are not steeped in drunkenness ; rural districts where hydrophobia is the universal blessing of humanity ; where the rain is distilled from the clouds and the spring bubbles from the rock, for the benefit of the brute creation alone ; where the infant draws swill milk from the maternal fount, and children are trained to tippling before they cut their teeth. Are such peoples examples for Americans ? Shall we tread in the footsteps of a starving peasantry, too miserably poor to rise above vinegar under the name of wine ? Shall we get up a wine-drinking paradise, and set our daughters to breaking stone on the roads, and harness our wives and mothers to carts and plows ? Is this the style of reformation from intemperance which our enthusiastic friend Ross Browne, the oracle of the wine makers, contemplates, when he urges that "every man, woman and child should be saturated like himself with California wine on the brain " ?

Besides, how is wine to exclude stronger drinks and to take their place, when the wine-growers themselves, in convention assembled, declare that "the success of the wine manufacture in Califor-

nia depends on brandy," and that a large portion
of the wine must be distilled into brandy in order
to carry out their project? They have already
succeeded in obtaining a reduction of the revenue
tax on grape brandy, and are preparing to flood
the market with that article.   We are sure to have
the curse of alcohol in all its shapes, under cover
of wine.*

"Must we then cut down all our vineyards?"
—is a common question which appeals to the
most sordid elements of the human character.
If a California lion were to make a vineyard his
lair, and to issue forth night by night, and year
by year, to prey on the adjoining farm-yards, and
if it were impossible to get rid of the enemy by
any other means than by the destruction of the
vineyard, what then? And if intemperance be the
beast of prey, and men women and children the
victims, does it weaken the analogy?

But there is no need of destroying the vineyards. We want grapes by the thousand tons as
a luxury and a food for the inhabitants. We want
the wholesome and delicious product on every
table on this coast, as a rich blessing, but not
transformed into a burning curse. By grafting
the common vines with choice varieties, by the
production of raisins, and by other devices which

---

* At the commencement of the wine manufacture several years ago,
various depositories in San Francisco exhibited the inscription on their
signs—"CALIFORNIA WINES." A year or two later the signs read:
"CALIFORNIA WINES AND BRANDY." Now they read: "CALIFORNIA
BRANDY AND WINES." Typical'

the sagacity of our people will develop, the vine-yards will be' a source of health and wealth, of comfort and good morals. The surplus and infe-rior product may even be converted into alcohol for legitimate use in the industrial arts, which consume an immense and an increasing quantity. "Where there is a will there is a way."

Many years ago the same outcry was made in regard to apple orchards, and in the interests of the cider makers. Orchardists have since discov-ered better and more profitable uses for their fruit, in supplying the home market, and that of the West Indies and of Great Britain. The trans-continental railroad will open a new market for our grapes.

If there is any doubt as to the effect of wine drinking at the present day, look at the records of the past, when no other intoxicating drink was in use—long, very long before the Arabian concen-trated the subtle poison by distillation. Histo-ry, sacred or profane, tells of no wine-drinking people that was not cursed with drunkenness. Scarcely were the waters of the deluge dried up, when the demon alighted on the new-born earth ; and the first print of his cloven foot was in the first vineyard. Read it, ye wine-bibbers, who, not content yourselves to imbibe, insist on cramming it down our throats and the throats of our chil-dren : read the history of Noah ! It appears to have been written expressly for you ! "AND NOAH PLANTED A VINEYARD ; AND HE DRANK WINE,

AND WAS DRUNKEN. " There is the whole story !
The wine must have been pure and genuine ;
equal at least to the best Los Angeles or Sonoma.
No shrewd Frenchman, no smart Yankee could
have fabricated or adulterated *that* wine.

Look at the consequences and learn a lesson !
If the notion be true that African slavery grew
out of the curse pronounced by the father of the
wine-culture, what a fiery deluge has come down
upon the nations from that first vintage, culmi-
nating in the greatest of rebellions, by which hun-
dreds of thousands of our brethren were consigned
to gory graves ! With course more deadly still
has that fiery stream swept downward through the
generations of Shem and Japhet, scathing and
burning and blasting the bodies and souls of men,
bursting through all the barriers of civilization
and Christianity, consigning to a life of torture
and a death of infamy tens of thousands, yea,
millions of men endowed by nature with highest
and noblest intellects, and crushing out all hap-
piness and hope from the domestic circle, from
the hearts of mothers, wives, and children, in city
and hamlet, in palace and hovel, until from every
habitable corner of the wide world—

> " Hoarse, horrible and strong
> Goes up to Heaven the agonizing cry,
> Piercing the hollow arches of the sky—
> How long, oh God ! how long !"

In the history of Lot the horrid crime of incest
is associated with the wine-cup. Amnon, the son

of David, was slain when his heart was merry with wine. Elah, king of Israel, was murdered by his servants while drinking himself drunk with wine. Benhadad and his thirty-two kings, whilst drinking themselves drunk with wine, were surprised and vanquished. The fat valley of the Ephraimites was as famous for its wines as our own Sonoma; and hearken, oh ye of California, to the dread sentence: "Wo to the crown of pride, to the drunkards of Ephraim, whose glorious beauty is a fading flower!" Was it not wine that caused priest and prophet to err? Was it not wine which swallowed them up, and made them stumble in judgment; and filled all their tables with vomit and filthiness? What was it we are warned not to look upon, red and sparkling in the cup? In what drink was lurking the bite of the serpent and the sting of the adder? Was it not wine which made wo and sorrow, contention and babbling, wounds and redness of eyes? Was not wine a mocker from the first? And has it not been a mocker always? And will it not be a mocker to us and to our children?

Read the history of Belshazzar! At midnight, in mirth and revelry, a thousand golden goblets overflowing with wines pure and costly as the power and wealth of Babylon could command, the ominous inscription blazed forth on the wall. That night Belshazzar was slain, and the pride and glory of Babylon departed. Would to Heaven every wine grower in California could read as a timely warning the same inscription in letters of fire, on every cask in his cellar, and on every wall in his dwelling!

It is worthy of note that total abstinence from all intoxicating drinks was marked by the favor of Heaven through all the ages of antiquity. The Nazarites took the temperance pledge in its most rigid form, and the condemnation of the Almighty was pronounced against those who offered them wine to drink—a condemnation which those in our day would do well to consider, who talk of forcing it down everybody's throat. The Rechabites went still further than the Nazarites and vowed abstinence for themselves and their posterity; and on that account they were blessed with the divine promise that they should always have a representative to stand before God. During their religious ministration the priests dare not taste wine. Samuel was sealed to a holy life by his mother's vow that he should be a teetotaler. The mother of Sampson never poisoned her blood with a drop of fermented grape juice, and her son grew up to be a prodigy of strength by drinking only of the crystal spring, God himself having administered the pledge. Hear Milton on this topic:

O, madness! to think use of strongest wines
And strongest drink our chief support of health,
When God, with these forbidden, made choice to rear
His mighty champion, strong above compare,
Whose drink was only from the limpid brook.

Profane history tells the same story of all wine-drinking nations A carousal on wine led to the assassination of Philip of Macedon. Alexander, educated to the wine cup, while drunk murdered his best friend ; and the conquerer of the world was himself at length conquered and slain by this "temperance" drink. The Scythians, in the vigor of a sober life, and before they had taken to wine, subjugated all Western Asia ; but they became wine-drinkers, and in less than thirty years their arms were unnerved and the sceptre fell from their grasp. Brennus and his victorious Gauls, after sweeping through Italy and taking possession of proud and haughty Rome, encamped amid the vineyards ; and there while steeped in drunken slumbers, fell ingloriously beneath the sword of Camillus. The Marsi, a tribe of Germans at war with Rome, in the stupor consequent on a wine feast, were slaughtered by the Romans without the loss of a single man. In like manner the Cambrians, triumphant and resistless until they descended from the Alps and reached the vineyards of Italy, there swallowed the seductive draught and fell an easy prey to the sword of Marius. Hannibal, after dissolving the Alps, in the figurative language of the historian, not with wine but with vinegar, poured down upon Italy like an avalanche and planted his victorious standards before the gates of Rome. But while he rested in the vineyards of Capua and held dalliance with the enchantress, the Romans despatched an army to Africa, and Carthage fell instead of Rome.

For ten long years did the brave and vigilant Trojans defend their noble city from the fleets and armies of Greece : but a single hour of wine and revelry sufficed to blot Troy from the face of the earth.

Through the later periods of history we trace the same debasing, animalizing, brutalizing influence of wine on the human character. The intemperance of England was born of wine. The marriage of Henry II with a French princess who possessed extensive vineyards in the south of France, was the means of flooding the kingdom with wine, and by this means, the historian remarks, "the land was filled with drink and drunkards." A few years prior to this, had occurred the melancholy shipwreck of the only son of Henry I, with his entire retinue. This accomplished prince set sail from France on his return to England, in a vessel with fifty rowers, with a bounteous supply of wine, pure, rich and sparkling. The seamen solicited the tempting draught, captain and helmsman and crew became inebriated, and the vessel went down, with the prince and his young sister, and one hundred and forty of the flower of England's nobility.

Another striking illustration occurred during the disastrous retreat of Sir John Moore in the Spanish peninsula, before the legions of Napoleon. The great wine vaults of Bembibre, says the historian, proved more fatal to the British army than the sword of the enemy. The rear guard did all the fighting, and in the midst of incredible hardships, made a glorious defense of the flying troops. And yet on their arrival at Corunna, the losses of the rear guard were less than of any other division of equal numbers in the army.

It may be said that other fermented drinks have been in use and are responsible for the drunkenness of the past; and that mixed and drugged wines are mentioned in the Bible and elsewhere as supplying the means of indulgence. But there is no fermented drink as strong in alcohol as the purest and richest wine, made from the grape. This luscious fruit exceeds all other fruits in the richness and sweetness of its juice, and of course in the quantity of poison which fermentation develops in it. I am aware that some writers mention palm wine, made from the juice of the palm tree, as more intoxicating than that of the grape. But this is not believed by good authorities. As for mixed wines, it is well known that in a large number of instances, the mixture referred to was simply with water, in accordance with a common custom of the ancients. And as for drugged wines, point me to the drug so poisonous to the moral nature of man as alcohol itself. Strychnia and prussic acid and arsenic will destroy life. But who ever heard of a man killing his friend while under the influence of strychnia? or a husband slaying his wife in consequence of swallowing prussic acid? or a father murdering his offspring because he had taken

arsenic? No, my friends ! you may ransack the
kingdoms of nature—you may analyze every mineral, every animal, every vegetable on the face of
the earth—you may torture the elements into all
sorts of chemical combinations—but you will find
nothing that God ever made, that man ever con-
trived, comparable to alcohol in its power to
transform human beings into brutes—into fiends
—into devils.

Remember that scarcely three centuries have
passed since distilled liquors came into common
use ; that prior to that date, for three thousand
years, embracing the long ages covered by ancient
history sacred and profane, the drunkenness of the
world was essentially on wine ; that all the de-
nunciations of Holy Writ against intemperance
are leveled at indulgence in wine. In the experience of nations it has ever been associated with
luxury, corruption, weakness, vice and downfall,
and in the individual character, with murder and
rapine, with incest and prostitution, with sorrow
and death. What else so poisons the heart and
hatches crime in its hidden chambers? What else
so stifles the divine instincts in the human bosom
and nerves the arm to deeds of blood? What else
so defiles the issues of life and stamps the progeny
with idiocy, and breeds monsters instead of men?
What else so brazens the cheek of beauty, and
dissipates the innocence and virtue of woman, and
gives her over to the wiles of the seducer?

From the day when Noah planted the first vine-

yard and drank wine and was drunken, history
has most faithfully repeated itself. It is repeat-
ing itself already in California, and will continue
to do so. They who plant vineyards and drink
wine will be drunken, and they will curse their
offspring. It is a law of nature, operative on individuals and on peoples. To all nations the wine
cup has proved the "cup of the wrath of God."

> " Nations melt
> From power's high pinnacle when they have felt
> Its sunshine for a while, and downward go
> Like lauwine loosened from the mountain's belt."

No one who studies the philosophy of habit
with special reference to intemperance will con-
ceive it possible to rear a generation of sober
people on wine, or any other kind of intoxicating
drink. Habit never grows backward—it is always
aggressive. Its conquests are permanent. It
never yields an inch. Whatever it gains it keeps
—and it grows from what it feeds on. One by
one its silken cords are wound about the limbs
till they form a chain stronger than adamant.
The idea of promoting temperance by wine-drinking is fraught with danger. If men could obtain
nothing but the weakest wines of the starving
peasantry of Europe, which have been compared
by travelers to vinegar and water, they would
doubtless be sober. But who dreams that such
a drink would be tolerated in California? Our
people are proverbial for wanting and getting the

best of everything. "Generous wine" will be necessary for their palates. And will they stop there? The boy tends to grow into the man not more surely than the appetite for wine tends to brandy.

Persons who study and understand the philosophy of habit, will appreciate the importance of this subject, especially to the young men of California. A glass of wine is a small thing. An occasional drink seems like a very innocent indulgence. But no man is safe who does not give heed to small things. Only by faithfulness in little things is self-government to be acquired. In no other way can we become rulers over more. To avoid beginnings affords the only security against the acquisition of bad habits. "It is the first step that costs." Samuel Johnson said to a lady who urged him to take a glass of wine on the plea that a little could do no harm—"Madam, it is easier to drink none than to drink a little!" His experience has been the experience of millions. On the margin of the whirlpool, the smoothly circling waters give no sign of the abysmal vortex to which they tend : nor is their apparent course toward the centre. The first glass— the first act of indulgence—takes the bark of life into that smoothly circling tide ; and as it sweeps gracefully around, every turn brings it nearer and nearer to the unseen destruction, and makes it harder and harder to escape. My young friend and brother ! let me enjoin thee to steer thy bark upon the safe and open sea of total abstinence, and to give a wide berth to the whirlpool !

The industrial and commercial aspect of the wine culture, as a source of wealth to individuals and the State, in the minds of many persons overshadows every other consideration. There are not many countries in the world where gold is more powerful than here. The love of gold has shipwrecked many a noble soul during the short life of our commonwealth. In political circles it has been said that every man has his price. The maxim has proved true in other circles. Gold is king in California. And shall it be God as well as king ? Oh, you, my friends, who are disposed to give due weight to moral considerations even when they come in conflict with your supposed pecuniary interests, or those of society at large —who look to the moral welfare of the present and coming generations as above all price—who believe that it profiteth not a man to gain the whole world and lose his own soul, and that all the vineyards of California are not worth as much as one human soul—you who feel disheartened at the mighty deluge which seems about to overwhelm us, and who are ready to regard all effort to stay the flood as vain and futile—let me entreat you to stand by your principles in the hope and faith that are born of a good cause! Compromise not a hair's breadth with right and duty! Remember that man proposes, but that a Power higher than man disposes! Bear in mind the recent, signal interposition of that Power in ridding us of one great national curse! Look back a few short years and contemplate the gloom and darkness which rested on our destiny—a gloom and darkness that no human foresight could penetrate! And behold our country now! its glorious sun shining with redoubled brightness! its national banner waving over thirty millions of freemen, and not a human chattel among them! our land, but yesterday prostrate and bleeding, crippled, crushed and exhausted, risen like a giant refreshed from slumber, and challenging the admiration of the nations as the Great Republic of the world! And beholding this, despair of nothing! Acquit yourselves manfully of every duty, and labor for that day, which in God's miraculous providence may be nearer at hand than priest or prophet can perceive, when that other national curse, which like the frogs and lice of Egypt has penetrated even to our dough troughs and bed chambers, shall meet the doom of slavery! Then shall earth ring out her deliverance and her liberty to Heaven, and Heaven echo back the glad response! Then shall the morning stars sing together, and the sons of God shout for joy!

Lightning Source UK Ltd.
Milton Keynes UK
UKHW020001280223
417750UK00005B/318